Synthetic Genomics - From BioBricks to Synthetic Genomes

Edited by Miguel Fernández-Niño and Luis H. Reyes

Published in London, United Kingdom

IntechOpen

Supporting open minds since 2005

Synthetic Genomics - From BioBricks to Synthetic Genomes
http://dx.doi.org/10.5772/intechopen.94713
Edited by Miguel Fernández-Niño and Luis H. Reyes

Contributors
Daniel A. Charlebois, Gábor Balázsi, Michael Tyler Guinn, Kevin S. Farquhar, Huafeng Lin, Haizhen Wang, Aimin Deng, Minjing Rong, Lei Ye, Lei Shi, Tuanmei Wang, Xiangwen Peng, Miguel Fernández-Niño, Angie Burgos-Toro, Martin Dippe, Andres Felipe Vásquez, Eric Pierschel, Ludger Aloisius Wessjohann, Vishnu Sharma, Tarun Kumar Kumawat, Anjali Pandit, Bhoomika Sharma, Pooja Agarwal, Bhawana Sharma, Preeti Nain, Manish Biyani, Luis H. Reyes

First published in London, United Kingdom, 2022 by IntechOpen
IntechOpen is the global imprint of INTECHOPEN LIMITED, registered in England and Wales, registration number: 11086078, 5 Princes Gate Court, London, SW7 2QJ, United Kingdom
Printed in Croatia

British Library Cataloguing-in-Publication Data
A catalogue record for this book is available from the British Library

Additional hard and PDF copies can be obtained from orders@intechopen.com

Synthetic Genomics - From BioBricks to Synthetic Genomes
Edited by Miguel Fernández-Niño and Luis H. Reyes
p. cm.
Print ISBN 978-1-83969-638-1
Online ISBN 978-1-83969-639-8
eBook (PDF) ISBN 978-1-83969-640-4

For EU product safety concerns:
IN TECH d.o.o., Prolaz Marije Krucifikse Kozulić 3, 51000 Rijeka, Croatia, info@intechopen.com or visit our website at intechopen.com.

Meet the editors

Dr. Miguel Fernández-Niño is a biologist from The National University of Colombia with a master's degree in Biochemistry from the same institution and a Ph.D. in Biochemical Engineering from Jacobs University, Bremen, Germany. He is currently working as a researcher at the Leibniz Institute of Plant Biochemistry, Halle (Saale), Germany. Dr. Fernández-Niño has extensive experience in synthetic biology, metabolic engineering, genome mining/editing, and molecular biotechnology, which can be applied to engineer and improve microorganisms with biotechnological applications.

Dr. Luis H. Reyes is a highly skilled and forward-thinking chemical engineer with a Ph.D. in Chemical Engineering and broad knowledge and experience in several fields of molecular and synthetic biology, genetics, evolutionary and protein engineering, and bioprocess engineering, using diverse microorganisms including *Escherichia coli*, *Saccharomyces cerevisiae*, *Pichia pastoris*, *Clostridium thermocellum*, *Kluyveromyces lactis,* and *Lactobacillus brevis*. Currently, he is an associate professor in the Department of Chemical and Food Engineering, Universidad de Los Andes, Bogota, Colombia.

Contents

Preface

Understanding complex biological systems and their relation to biological function is a current challenge in biological sciences. During the last two decades, the field of synthetic biology has allowed the design and construction of synthetic biological systems at different scales of complexity. The first level of complexity to be explored was related to the design of BioBricks (e.g., promoters, coding sequences, terminators, reporters, and regulatory sequences), which are the essential building blocks for pathway engineering. Over the last decade, an astonishing improvement in sequencing technologies has resulted in millions of bits of information in public databases. This is a valuable mine of data for identifying novel BioBricks so far not elucidated, many of them attributed to novel biological functions.

Moreover, the development of low-cost DNA synthesis technologies has boosted the rational design of these BioBricks and their subsequent integration into synthetic gene networks. Many methodological approaches have been designed to simplify cloning and BioBricks' assembly into networks. This represents the second level of complexity, where synthetic metabolic networks can be designed, constructed, and studied as a modular part of a more complex biological system. Recently, different attempts to integrate these modules into complex systems have opened the possibility to engineer and construct entire synthetic genomes, representing a tremendous advance in Synthetic Biology. Accordingly, a new research field has emerged called synthetic genomics. The general goal of this novel area of synthetic biology is to engineer synthetic genomes from scratch using a set of pre-designed building blocks coupled in a hierarchical and modular way.

This book discusses the state of the art in synthetic genomics by presenting relevant examples in this emerging area. The introductory chapter, "From BioBricks to Synthetic Genomes," describes the evolution of synthetic genomics, starting from BioBricks design and ending with the construction of complete viral and bacterial synthetic genomes. "Multi-*Omics* Data Mining: A Novel Tool for BioBrick Design" provides a deeper explanation of the principles of BioBricks design. "Applications of CRISPR/Cas Technology to Research the Synthetic Genomics of Yeast" and "CRISPR-Cas9: Role in Processing of Modular Metabolic Engineered Bio-Based Products" present examples of technologies employed to engineering genomes using CRISPR-based approaches. The final chapter, "Synthetic Gene Circuits for Antimicrobial Resistance and Cancer Research," describes the use of mathematical modeling for genome engineering. This book is a valuable tool for students and scientists who would like to become familiar with this new research area.

We would like to thank the researchers from the Group of Product and Process Design at Universidad de Los Andes, Colombia, and the Group of Biotechnology at the Leibniz Institute of Plant Biochemistry, Germany, for their contributions, correct observations, and excellent knowledge that improved this work. We would like to especially thank Dr. Prof. Ludger Wessjohann, Dr. Prof. Luis H. Reyes, and Dr. Martin Dippe for their valuable contributions and fruitful discussions. We would also like to thank Author Service Managers Iva Ribic and Marijana Francetic at IntechOpen for their support and contributions during the editorial process.

Miguel Fernández-Niño
Researcher at the Department of Bioorganic Chemistry,
Leibniz Institute of Plant Biochemistry,
Halle, Germany

Luis Humberto Reyes
Associate Professor at the Department of Chemical and Food Engineering,
Universidad de Los Andes,
Bogotá, Colombia

Section 1

Rational Design of BioBricks for Genome Assembly

Chapter 1

Introductory Chapter: From BioBricks to Synthetic Genomes

Luis Humberto Reyes and Miguel Fernández-Niño

1. Introduction

One of the goals of Synthetic Biology is to design novel biological systems by the rational assembly of biological parts (BioBricks) into artificial metabolic networks [1–4]. These engineered networks can be further coupled to create highly complex biological systems [3]. Such systems are arranged in living organisms that work as biological chassis to hold and express the engineer networks and produce a desired phenotype [4–7]. Several technologies have been developed during the last 20 years to improve our ability to engineer novel biological systems. This has been observed at different levels of complexity, including the development of technologies for the identification, design, and synthesis of BioBricks and the expansion of innovative protocols for pathway assembly/modeling and genome-editing technologies in the fine-tuning of biological systems (**Figure 1**).

Currently, BioBricks can be identified (mined) from the large amount of information contained in public databases (e.g., BRENDA [8], GenBank [9], PANTHER [10], UniProt [11], etc.) or by selecting pre-designed BioBricks from specialized databases such as the iGEM Parts Registry [12] and the BioMaster DataBase [13]. These engineered BioBricks can be obtained by traditional methods, including PCR [14], or by using cutting-edge technologies such as the *de novo* synthesis of large fragments of nucleic acids (synthetic genes [15]), whose prices have been remarkably decreasing during the last years. The selected BioBricks can be further engineered to optimize their heterologous expression and ease subsequent assembly, expression, and purification [2, 16]. Different protein tags have been designed for this purpose, including solubility and affinity tags and tags aimed to simplify subsequent cloning in expression vectors [17]. Once the desired BioBricks have been obtained, they are usually coupled to other BioBricks to create artificial networks of higher complexity [3, 18]. Several technologies have been developed to assemble multiple BioBricks in artificial networks, including PCR-dependent cloning [19], seamless cloning [20], recombinational approaches [21], among others.

The genetic expression of these engineered artificial networks can be further optimized by modifying the network topology (e.g., changing from operon to monocistronic topologies [20]) or by adding regulatory elements such as feedback loops, oscillators, riboswitches, and protein scaffolds [3]. The behavior of these artificial networks and their regulatory elements can be studied through *in silico* modeling to predict the most appropriate topology for the artificial system to be designed. Several tools and software for *in silico* modeling of metabolic pathways have been designed to determine the effect of the expression of a particular artificial network on the global metabolic response of the host organism [22–26]. For example, it is currently possible to identify genes in the host organism through *in silico* modeling, whose deletion may result in a higher expression of the desired phenotype or even predict the most efficient set of reactions required to produce valuable compounds

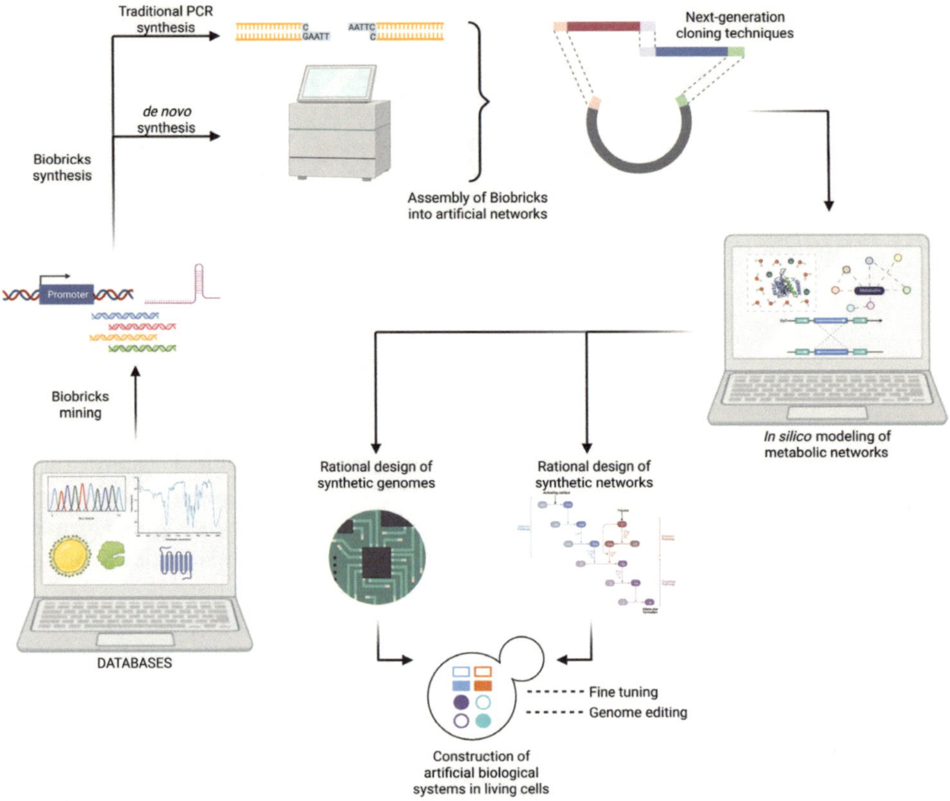

Figure 1.
Frequently used steps to engineer artificial biological systems and synthetic genomes using BioBricks as building blocks.

using reconstructed metabolic models [24, 25]. The predictive ability of these models relies on our understanding of the genome, transcriptome, and the global metabolism of the selected host. Thus, it is not surprising that most of the host organisms used in synthetic biology are widespread model organisms such as well-characterized strains of *Escherichia coli* and *Saccharomyces cerevisiae*, for instance. These organisms have been studied for generations, and different toolboxes for their metabolic engineering/genome editing have been previously designed [27].

Interestingly, with the reduction of sequencing prices and the development of novel methodologies for long-read sequencing (e.g., Oxford Nanopore [28] and PacBio technologies [29]), the genome of a large number of new (so far not characterized) organisms have been recently elucidated. This has been concurrent with the development of new technologies for the identification, characterization, and quantification of metabolites, proteins, and lipids using last-generation liquid and gas chromatographic columns coupled to mass spectrometry (LC-MS and GC-MS) and the development of new devices for nuclear magnetic resonance (NMR) analytics [30–32]. In addition, novel technologies for improving genome editing such as CRISPR-Cas9 have opened the possibility to expand our ability to engineer novel biological systems in living cells, as never before [33, 34]. Thus, the opportunity to design and construct an entire genome is now a reality with the current technological advances. This has opened a new field of research (known as Synthetic Genomics [35]) to engineer and assemble entire artificial genomes or larger parts of genomes in living organisms by using the principles of synthetic biology previously summarized. A genome is considered synthetic if the building blocks used for its assembly were originated by chemical synthesis [36].

2. Current advances in Synthetic Genomics

"*One needs to write synthetic DNA sequences in order to better understand the grammar of life*"-with this sentence, Schindler and co-workers, in their 2018 review, accurately summarized the primary motivation behind the fast progress of synthetic genomics [36]. Accordingly, one of the main goals of synthetic genomics is to improve our understanding of genome fundamentals [35, 36]. Thus, the construction of entire synthetic genomes has allowed the study of their constitutive building blocks, considering the effect of the whole biological system on biological function.

As previously described in this chapter, synthetic genomes are engineered following a hierarchical and modular assembly starting from synthetic genes, gene clusters, artificial metabolic pathways, and chromosomes. Currently, two approaches can be utilized to assemble a synthetic genome into an organism: (1) using a heterologous host or (2) using a native host as a chassis for chromosome replacement [35, 36]. Heterologous hosts are well-known model organisms with an extensive toolbox for genetic engineering that simplify the subsequent assembly (e.g., *E. coli* and *S. cerevisiae*). However, their engineering capacity might be restricted by the size and number of synthetic chromosomes to be assembled. On the other hand, native hosts are advantageous for chromosome replacement. Still, most of them are not well-characterized organisms, or there is a lack of tools for their genetic engineering. Currently, many native microorganisms have been characterized in response to the fast development of sequencing and genome editing techniques as previously described [27–29].

Most of the synthetic genomes that have been successfully assembled are viral and bacterial, with a smaller genome size than eukaryotes. For example, the viral genome of the Poliovirus (7.5 kb size) was entirely synthesized 20 years ago [37] with the technology available at the time. More than 15 years later, a fully synthetic genome was assembled for the Horsepox virus (212 kb size) using the latest technology, which allowed the assembly of a synthetic genome that is more than 20 times bigger as compared to Polio genome size [38]. Remarkably, Thao et al. (2020) have recently engineered and assembled the entire genome of the virus Severe Acute Respiratory Syndrome Coronavirus 2 (SARS-CoV-2) responsible for the current pandemic disease (COVID-19) using an *S. cerevisiae* platform [39]. This heterologous platform for the synthetic assembly of viral genomes constitutes a valuable tool to reconstruct different RNA viruses (from different families) in a short time using pre-designed synthetic building blocks [39]. Regarding the reconstruction of synthetic bacterial genomes, most of the research has been focused on species from the genus Mycoplasma with genome sizes ranging from 583 kb (in *Mycoplasma genitalium*) to 1079 kb (in *Mycoplasma mycoides*) [40, 41]. It is important to mention that up to now, there are no eukaryotic genomes that have been entirely reconstructed. However, there are still several efforts to engineer eukaryotic genomes, such as the Sc2.0 international project (Synthetic Yeast), aimed to assemble the world first eukaryotic synthetic genome from scratch and the Genome Project-write (GP-write) that is an international project aimed to reconstruct the entire genome of a large number of cell lines within the next years.

Currently, the research in synthetic genomics has moved one step forward to design new genomes that are different from the ones found in nature [35, 36]. Thus, the synthetic genome of *M. mycoides* has been reduced in size by 49.3%, which resulted in a new platform to discover new biological functions [42]. Similarly, a synthetic *E. coli* genome is currently being developed [43] to remove/replace codons in the codon sequence that allow the expression of proteins with non-natural amino acids. These examples show the relevance of synthetic genomics as a promising area to explore novel biological functions using completely unnatural biological systems. The research in this area is expected to increase in response to developing novel technologies and bioinformatics tools to rational design and analyze complex biological systems. Consequently, new platforms must be characterized to increase the number of organisms used as a host for chromosome assembly. New bioinformatics tools (Genomic design software) must still be designed to predict and study the behavior of the engineered synthetic genomes. Finally, it is important to mention that this emerging area has not only the potential to boost science but also can be used for harmful purposes. Consequently, a deeper discussion on synthetic genomics' ethical, social, and ecological consequences must still be conducted with scientists, politicians, and communities.

3. Impact and risks

The intentional or accidental release of genetically modified organisms into the environment could have significant negative impacts on both human and environmental health. This biological revolution, together with advances in biotechnology, could be used to improve the biological properties of viruses simply by altering resistance to antiviral agents, modifying antigenic properties, modifying the tropism, pathogenesis, and transmissibility of tissues, "humanizing" zoonotic viruses, and creating designer super-pathogens. The main paradigm shift may be that the approach is less technically demanding and more design-based, requiring only limited technical expertise because the genome can be synthesized and purchased from commercial vendors, government-sponsored facilities, or from rogue basement operations (e.g., bioterrorist sponsored organizations or private entrepreneur). The main technical support could include a competent research technician and minimal equipment to isolate recombinant pathogens from recombinant DNAs.

These potential impacts require governance methods and research guidelines that promote their ethical and responsible use. Under the precautionary principle, a rigorous risk assessment and inclusion of diverse stakeholder perspectives should be applied in the development and management of innovative synthetic biology applications and products. The precautionary principle states that when human activities can lead to unacceptable harm that is scientifically plausible but uncertain, steps must be taken to avoid or lessen that harm.

Author details

Luis Humberto Reyes[1] and Miguel Fernández-Niño[2]*

1 Grupo de Diseño de Productos y procesos (GDPP), Departamento de Ingeniería Química y de Alimentos, Universidad de los Andes, Bogotá, Colombia

2 Department of Bioorganic Chemistry, Leibniz-Institute of Plant Biochemistry, Halle, Germany

*Address all correspondence to: mfernand@ipb-halle.de

References

[1] Leggieri PA, Liu Y, Hayes M, Connors B, Seppälä S, Malley MAO, et al. Integrating systems and synthetic biology to understand and engineer microbiomes. Annual Review of Biomedical Engineering. 2021;**23**:169-201

[2] Hughes RA, Ellington AD. Synthetic DNA synthesis and assembly: Putting the synthetic in synthetic biology. Cold Spring Harbor Perspectives in Biology. 2017;**9**:a023812

[3] Agapakis C, Silver P. Synthetic biology: Exploring and exploiting genetic modularity through the design of novel biological networks. Molecular BioSystems. 2009;**5**:704-713. DOI: 10.1039/b901484e

[4] Singh V. Recent advancements in synthetic biology: Current status and challenges. Gene. 2014;**535**:1-11. DOI: 10.1016/j.gene.2013.11.025

[5] Calero P, Nikel PI. Minireview chasing bacterial chassis for metabolic engineering: A perspective review from classical to non-traditional microorganisms. Microbial Biotechnology. 2019;**12**:98-124. DOI: 10.1111/1751-7915.13292

[6] Yu D, Wang M, Zhu G, Ge B, Liu S, Huang F. Enhanced photocurrent production by bio-dyes of photosynthetic macromolecules on designed TiO_2 film. Scientific Reports. 2015;**5**:9375

[7] Liu J, Wu X, Yao M, Xiao W. Chassis engineering for microbial production of chemicals: From natural microbes to synthetic organisms. Current Opinion in Biotechnology. 2020;**66**:105-112. DOI: 10.1016/j.copbio.2020.06.013

[8] Chang A, Jeske L, Ulbrich S, Hofmann J, Koblitz J, Schomburg I, et al. BRENDA, the ELIXIR core data resource in 2021: New developments and updates. Nucleic Acids Research. 2021;**49**:498-508. DOI: 10.1093/nar/gkaa1025

[9] Clark K, Karsch-Mizrachi I, Lipman DJ, Ostell J, Sayers EW. GenBank. Nucleic Acids Research. 2016;**44**:67-72. DOI: 10.1093/nar/gkv1276

[10] Mi H, Ebert D, Muruganujan A, Mills C, Albou L, Mushayamaha T, et al. PANTHER version 16: A revised family classification, tree-based classification tool, enhancer regions and extensive API. Nucleic Acids Research. 2021;**49**:394-403. DOI: 10.1093/nar/gkaa1106

[11] The UniProt Consortium. UniProt: The universal protein knowledgebase in 2021. Nucleic Acids Research. 2021;**49**:480-489. DOI: 10.1093/nar/gkaa1100

[12] Smolke CD. Building outside of the box: iGEM and the BioBricks foundation. Nature Biotechnology. 2009;**27**:1099-1102. DOI: 10.1038/nbt1209-1099

[13] Wang B, Yang H, Sun J, Dou C, Huang J, Charles TC. BioMaster: An integrated database and analytic platform to provide comprehensive information about BioBrick parts. Frontiers in Microbiology. 2021;**12**:1-6. DOI: 10.3389/fmicb.2021.593979

[14] Shetty RP, Endy D, Knight TF Jr. Engineering BioBrick vectors from BioBrick parts. Journal of Biological Engineering. 2008;**12**:1-12. DOI: 10.1186/1754-1611-2-5

[15] Kosuri S, Church GM. Large-scale de novo DNA synthesis: Technologies and applications. Nature Methods. 2014;**11**:499-507. DOI: 10.1038/nmeth.2918

[16] Elena C, Ravasi P, Castelli ME, Peirú S, Menzella HG. Expression of codon optimized genes in microbial

systems: Current industrial applications and perspectives. Frontiers in Microbiology. 2014;**5**:1-8. DOI: 10.3389/fmicb.2014.00021

[17] Kimple ME, Brill AL, Pasker RL. Overview of affinity tags for protein purification. Current Protocols in Protein Science. 2013;**73**:1-23. DOI: 10.1002/0471140864.ps0909s73

[18] Young R, Haines M, Storch M, Freemont PS. Combinatorial metabolic pathway assembly approaches and toolkits for modular assembly. Metabolic Engineering. 2021;**63**:81-101. DOI: 10.1016/j.ymben.2020.12.001

[19] Tillett D, Neilan BA. Enzyme-free cloning: A rapid method to clone PCR products independent of vector restriction enzyme sites. Nucleic Acids Research. 1999;**27**:26-28

[20] Xu P, Vansiri A, Bhan N. ePathBrick: A synthetic biology platform for engineering metabolic pathways in *E. coli*. ACS Synthetic Biology. 2012;**7**:256-266

[21] Park J, Throop AL, Labaer J. Site-specific recombinational cloning using gateway and in-fusion cloning schemes. Current Protocols in Molecular Biology. 2015;**110**:1-23. DOI: 10.1002/0471142727.mb0320s110

[22] Kumar VS, Maranas CD. GrowMatch: An automated method for reconciling in silico/in vivo growth predictions. PLoS Computational Biology. 2009;**5**:18-20. DOI: 10.1371/journal.pcbi.1000308

[23] Starcevic A, Zucko J, Simunkovic J, Long PF, Cullum J, Hranueli D. ClustScan: An integrated program package for the semi-automatic annotation of modular biosynthetic gene clusters and in silico prediction of novel chemical structures. Nucleic Acids Research. 2008;**36**:6882-6892. DOI: 10.1093/nar/gkn685

[24] Pharkya P, Burgard AP, Maranas CD. OptStrain: A computational framework for redesign of microbial production systems. Genome Research. 2004;**14**:2367-2376. DOI: 10.1101/gr.2872004

[25] Burgard AP, Pharkya P, Maranas CD. OptKnock: A bilevel programming framework for identifying gene knockout strategies for microbial strain optimization. Biotechnology and Bioengineering. 2003;**84**(6):647-657. DOI: 10.1002/bit.10803

[26] Fong SS, Palsson BØ, Herrga MJ. Identification of genome-scale metabolic network models using experimentally measured flux profiles. PLoS Computational Biology. 2006;**2**:e72. DOI: 10.1371/journal.pcbi.0020072

[27] Redden H, Morse N, Alper HS. The synthetic biology toolbox for tuning gene expression in yeast. FEMS Yeast Research. 2015;**15**:1-10. DOI: 10.1111/1567-1364.12188

[28] Jain M, Olsen HE, Paten B, Akeson M. The Oxford nanopore MinION: Delivery of nanopore sequencing to the genomics community. Genome Biology. 2016;**17**:1-11. DOI: 10.1186/s13059-016-1103-0

[29] Xie H, Yang C, Sun Y, Igarashi Y, Jin T, Luo F. PacBio long reads improve metagenomic assemblies, gene catalogs, and genome binning. Frontiers in Genetics. 2020;**11**:1077. DOI: 10.3389/FGENE.2020.516269/BIBTEX

[30] Beale DJ, Pinu FR, Kouremenos KA, Poojary MM, Narayana VK, Boughton BA, et al. Review of recent developments in GC-MS approaches to metabolomics-based research. Metabolomics. 2018;**14**:1-31. DOI: 10.1007/s11306-018-1449-2

[31] Wang Y, Hui S, Wondisford FE, Su X. Utilizing tandem mass

spectrometry for metabolic flux analysis. Laboratory Investigation. 2020;**101**:423-429. DOI: 10.1038/s41374-020-00488-z

[32] Li D, Gaquerel E. Next-generation mass spectrometry metabolomics revives the functional analysis of plant metabolic diversity. Annual Review of Plant Biology. 2021;**72**:867-891

[33] Gupta RM, Musunuru K. Expanding the genetic editing tool kit: ZFNs, TALENs, and CRISPR-Cas9. The Journal of Clinical Investigation. 2014;**124**:4154-4161. DOI: 10.1172/JCI72992. transcription

[34] Nidhi S, Anand U, Oleksak P, Tripathi P, Lal JA, Thomas G, et al. Novel CRISPR-Cas systems: An updated review of the current achievements, applications, and future research perspectives. International Journal of Molecular Sciences. 2021;**22**:1-42

[35] Wang L, Jiang S, Chen C, He W, Wu X, Wang F, et al. Synthetic genomics: From DNA synthesis to genome design. Angewandte Chemie (International Ed. in English). 2018;**57**:1748-1756. DOI: 10.1002/ANIE.201708741

[36] Schindler D, Dai J, Cai Y. Synthetic genomics: A new venture to dissect genome fundamentals and engineer new functions. Current Opinion in Chemical Biology. 2018;**46**:56-62. DOI: 10.1016/J.CBPA.2018.04.002

[37] Cello J, Paul AV, Wimmer E. Chemical synthesis of poliovirus cDNA: Generation of infectious virus in the absence of natural template. Science. 2002;**297**:1016-1018. DOI: 10.1126/SCIENCE.1072266

[38] Koblentz GD. The de novo synthesis of horsepox virus: Implications for biosecurity and recommendations for preventing the reemergence of smallpox. Health Security. 2017;**15**: 620-628. DOI: 10.1089/HS.2017.0061

[39] Thao TTN, Labroussaa F, Ebert N, V'kovski P, Stalder H, Portmann J, et al. Rapid reconstruction of SARS-CoV-2 using a synthetic genomics platform. Nature. 2020;**582**:561-565. DOI: 10.1038/s41586-020-2294-9

[40] Gibson DG, Benders GA, Andrews-Pfannkoch C, Denisova EA, Baden-Tillson H, Zaveri J, et al. Complete chemical synthesis, assembly, and cloning of a *Mycoplasma genitalium* genome. Science. 2008;**319**:1215-1220. DOI: 10.1126/SCIENCE.1151721

[41] Gibson DG, Glass JI, Lartigue C, Noskov VN, Chuang RY, Algire MA, et al. Creation of a bacterial cell controlled by a chemically synthesized genome. Science. 2010;**329**:52-56. DOI: 10.1126/SCIENCE.1190719

[42] Hutchison CA, Chuang RY, Noskov VN, Assad-Garcia N, Deerinck TJ, Ellisman MH, et al. Design and synthesis of a minimal bacterial genome. Science. 2016;**351**:aad6253. DOI: 10.1126/SCIENCE.AAD6253

[43] Ostrov N, Landon M, Guell M, Kuznetsov G, Teramoto J, Cervantes N, et al. Design, synthesis, and testing toward a 57-codon genome. Science. 2016;**353**:819-822. DOI: 10.1126/SCIENCE.AAF3639

Chapter 2

Multi-*Omics* Data Mining: A Novel Tool for BioBrick Design

Angie Burgos-Toro, Martin Dippe, Andres Felipe Vásquez, Eric Pierschel, Ludger Aloisius Wessjohann and Miguel Fernández-Niño

Abstract

Currently, billions of nucleotide and amino acid sequences accumulate in free-access databases as a result of the *omics* revolution, the improvement in sequencing technologies, and the systematic storage of shotgun sequencing data from a large and diverse number of organisms. In this chapter, multi-*omics* data mining approaches will be discussed as a novel tool for the identification and characterization of novel DNA sequences encoding elementary parts of complex biological systems (BioBricks) using *omics* libraries. Multi-*omics* data mining opens up the possibility to identify novel unknown sequences from free-access databases. It also provides an excellent platform for the identification and design of novel BioBricks by using previously well-characterized biological bricks as scaffolds for homology searching and BioBrick design. In this chapter, the most recent mining approaches will be discussed, and several examples will be presented to highlight its relevance as a novel tool for synthetic biology.

Keywords: genome, transcriptome, proteome, data mining, metabolic pathway, BioBricks design, multi-*omics*, synthetic biology

1. Introduction

1.1 The omics revolution

Within the last decades, a magnificent transformation in biology took place when a huge success in sequencing, bioinformatics, and bioanalytics was achieved. Several technologies were created to decrypt the metabolism of cells or interactions within tissues, organisms, and even entire ecosystems based on the identification of genes (genomics), mRNA (transcriptomics), proteins (proteomics), and metabolites (metabolomics) [1]. Since the discovery of the DNA structure by Watson and Crick in 1953 [2], an ever-increasing number of technologies for gene identification and characterization was established. One of the most relevant breakthroughs in DNA characterization was the invention of Sanger's sequencing in 1977 [3]. This sequencing technique uses chemical analogs of the deoxyribonucleotides (dNTPs, monomers of DNA strands) called dideoxynucleotides (ddNTPs), which lack the 3′ hydroxyl group that is required for extension of DNA chains and therefore cannot form a bond with the 5′ phosphate of the next dNTP [4]. The overall advantages of accuracy, robustness, and ease of use against other established methods led Sanger sequencing to become one of the most common technologies used to sequence DNA. Several improvements were subsequently applied to this technique, such as the use of fluorometric detection and capillary-based electrophoresis, thus contributing to the development of automated DNA sequencing machines [5–11]. These machines allowed researchers to obtain sequence reads slightly less than one kilobase (kb) in length and boosted the development of other crucial technologies such as the Polymerase Chain Reaction (PCR) in 1985 and the recombinant DNA technology in the following years [12, 13].

In parallel to the development of large-scale dideoxy sequencing methods, a new technique set the novum for next-generation DNA sequencers. This approach remarkedly varies from the abovementioned methods as it does not involve the use of radio- or fluorescently labeled dNTPs. Instead, it is based on a luminescent method for measuring pyrophosphate synthesis in a process called pyrosequencing [14]. This sequencing technology is a two-enzyme process starting with the conversion of pyrophosphate into ATP (by an ATP sulfurylase) and the subsequent use of ATP as a substrate for luciferase, thus emitting light proportional to the amount of pyrophosphate available. Pyrosequencing became a popular technique for two major reasons: (i) it uses natural nucleotides instead of modified ones, and (ii) that sequencing results can be obtained in real-time without requiring time-consuming electrophoresis. In addition to pyrosequencing, other sequencing technologies were also devolved - the most important probably being the Solexa method, later acquired by the company Illumina [15]. Hereby, adapter-bracketed DNA molecules pass a lawn of complementary oligonucleotides bound to a flow cell. This method involves solid-phase PCR with neighboring clusters of clonal DNA strands in a process called "bridge amplification" [15–17]. Apart from Illumina, which is probably the most important technique currently in use, other sequencing companies established their novel methodologies [18, 19], which are known as the second-generation sequencing techniques. The most notable second-generation sequencing platform is probably Ion Torrent. It is the first "post-light sequencing" technology with neither using fluorescence nor luminescence. Its methodology is based on beads bearing clonal populations of DNA fragments washed over a pico well plate, thereby releasing protons measured via the generated pH difference [20].

Recently, a third sequencing generation started with the invention of S. Quake in 2003 termed Single Molecule Sequencing (SMS) [21, 22]. Its principle is similar to Illumina but skipping bridge amplification. In SMS, DNA templates attached to a planar surface and propriety fluorescent reversible terminator dNTPs (dubbed as "virtual terminators") are washed over one base at a time and imaged, before cleavage and cycling the adjacent base over. SMS has been recently improved in the Single-Molecule Real-Time (SMRT) platform from Pacific Biosciences, available for the PacBio machines [23]. During SMRT runs, DNA polymerization happens in arrays of microfabricated nanostructures called zero-mode waveguides (ZMWs) which are essentially tiny holes in a metallic film covering a chip. It allows visualization of single fluorophore molecules because the zone of laser excitation is so small that it allows distinction over the background of neighboring molecules in the solution [24]. Nonetheless, the probably most anticipated third-generation DNA sequencing method is nanopore sequencing which enables researchers to detect and quantify all types of biological molecules [25]. Its principle was theoretically established even before second-generation sequencing emerged by demonstrating that single-stranded RNA or DNA could be driven across a lipid bilayer through a large α-hemolysin ion channel by electrophoresis. Furthermore, passage through the channel blocks ion flow, decreasing the current for a length of time proportional to the length of the nucleic acid [26]. With Oxford Nanopore Technologies (ONT) as the first provider of nanopore sequencers and their nanopore platforms GridION and MinION [27, 28], the latter of which is a small, mobile phone-sized USB device (released in 2014) [29]. Despite the admittedly poor quality profiles currently observed, it is hoped that such sequencers represent a genuinely disruptive technology in the DNA sequencing field in the future, producing incredibly long read (non-amplified) sequence data far cheaper and faster than what was previously possible [28, 30]. The average read length, error rate, total number of reads, and run prices vary significantly among the different sequencing methodologies. Thus, the selection of the appropriate technology for sequencing is a crucial step that depends on the purpose of the study. For instance, Illumina and Ion Torrent produce accurate short reads ideal for the analysis of fragmented DNA, while PacBio and MinION produce long reads with a lower accuracy but very useful, for example, for the assembly of scaffolds during genome sequencing.

Similar to the development of advanced techniques for sequencing nucleic acids, other methods have been extensively developed for dissecting the proteome [31] and metabolome [32] of a multitude of organisms. Of these omics approaches metabolomics, however, is distinct from the others. In metabolomics not a set of linear (1D) molecules with a sequence of defined monomers (4 bases or 21 amino acids) is to be determined, but a wild bunch of different 3D compounds. Eventually, a large number of databases have been developed to collect all these information, which provide excellent platforms for data mining as will be discussed in the following chapters.

2. Genome and transcriptome data mining

The exponential accumulation of data in genomic databases during the last decades has motivated the creation of bioinformatics tools to explore, relate and understand the genetic information from a vast number of organisms [33, 34]. These bioinformatics tools have been validated by experimental data, thus strengthening the design and assembly of novel biological entities (i.e., genes, RNA molecules, proteins, and metabolites). Those biological entities that can be used as building blocks for the assembly of artificial biosynthetic pathways are known as BioBricks. Consequently, the selection and design of BioBricks is important to further create and understand complex biological systems and biofactories of relevance in industrial biotechnology [35]. The general idea of comparing genomic sequences to identify such novel components of different metabolic pathways is not new. In fact, early in the 1970s, several efforts were performed to elucidate physiological and metabolic information through the comparative analysis of genetic sequences [36–38]. Classical genetics and reverse genetics approaches were then used to identify, annotate, compare, and connect genetic clusters associated with biosynthesis, using previously reported genetic data sets [39, 40].

It was not until 1999 that Genome Mining (GM) formally emerged as a strategy for the computational analysis of genetic sequences that sought to recognize patterns between them within the framework of the human genome project. Later, alongside bioinformatics advances in the area of microbiology, GM acquires new attributes, building the concept known today: a bioinformatics approach that aims to predict DNA sequences associated with physiological and/or metabolic events, allowing the elucidation/prediction of metabolic pathways that lead to secondary metabolites of scientific and industrial interest [35, 38, 41, 42]. Today, GM is not limited only to genomic predictions but seeks a holistic approach that includes the entire spectrum of molecular biology, articulating the prediction of the products of gene expression, the control of that expression, as well as the identity and structure of those potential metabolites, strengthening the creation of biological models that allow the comparison, understanding, and manipulation of cellular molecular systems [41, 43].

GM was initially developed in bacterial models and demonstrated a high relevance for synthetic biologists and metabolic engineers, thus becoming one of the biggest breakthroughs in molecular biology and biotechnology [38, 44]. Between the 1990s and 2000s, the genus *Streptomyces* (which is well known for its production of valuable antibiotics) was extensively studied at the experimental level, which allowed the identification of a large number of gene sequences involved in secondary metabolite production, regulation and antibiotic resistance. Comparison of gene sequences between different species of this genus, revealed a total of about 30 Biosynthetic Gene Clusters (BGCs) associated with the biosynthesis of such secondary metabolites [45, 46]. Following these advances, GM was extended to study novel bacterial genera with abundant genomic information and was initially used to fight against bacterial resistance [47, 48]. During the last years, GM was successfully used as a tool for the identification of alternative pathways for the biosynthesis of different natural products in diverse microorganisms [33, 49], an approach which usually proved to be more efficient than other screening methods used for the identification of novel enzymes of relevance for the biosynthesis of secondary metabolites [33, 49].

Recently, GM was also scaled up to eukaryotic models, thus revealing that multiple BGCs contain not only relevant information regarding the biosynthesis of secondary metabolites but also valuable information to study evolutionary events and ecological adaptation of different gene clusters [38, 50, 51]. A good example of the vast collection of BGCs predicted up to now can be found on the "Atlas of Biosynthetic Gene Clusters", a database of the Joint Genome Institute founded in 2015. This Atlas contains data on predicted and experimental gene clusters related to many secondary metabolites. As of June 2021, there are a total of 411,006 biosynthetic gene clusters reported, of which only 1285 have been experimentally validated [52]. GM is completely dependent on bioinformatics and computational technology available for the analysis of a large dataset. Thus, to boost the potential of this information, the development of novel computational tools and algorithms as well as the interest of researchers to join this effort is still required [42, 51]. There are currently a variety of methods for performing GM using the available genomic information that will be further discussed hereafter.

2.1 Classical genome mining

The "classical" form of GM consists of the search for enzymes linked to the synthesis of secondary metabolites, by mining highly conserved sequences [35]. Before the current databases (composed of hundreds of genomic datasets and several bioinformatics tools) were established, novel sequences were evaluated by using reverse genetics, where genomic libraries were scanned for basic biosynthetic genes associated with a metabolic pathway of interest [38, 53]. Those annotations had to be performed manually and by obtaining experimentally corroborated results. This formed the basis of classical GM, which provided the first consensus sequences to be compared with the vast amount of novel sequences obtained from different next-generation sequencing platforms [54]. Both, reverse genetics and GM follow the same mining pattern: one or several reference sequences, whose enzymatic products were already experimentally validated, are used to compare them with the genomes of interest and to identify homologous sequences in the organism of interest. Sequences of interest are considered as being generally associated with catalytic domains and highly conserved motifs [35, 38].

Classical GM was initially focused on the identification of genomic clusters associated with enzymes for the production of secondary metabolites, that involve the following bacterial groups of enzymes and bioactive peptides: (i) polyketide synthases (PKSs); (ii) non-ribosomal peptide synthetases (NRPSs); and ribosomally and post-translationally modified peptides (RiPPs) [55–57]. Sequence comparison of these groups of proteins allowed the subsequent identification of conserved motifs that are currently helping to identify novel BGCs in pre-existing genomes, without resorting to the strenuous processes of experimentation and first considering the bioinformatic *in silico* approach [58]. Thus, numerous examples have demonstrated the advantage of GM as a successful screening tool for evaluating the ability of one organism to produce a particular metabolite based on the available BGCs information [59–61]. An example of this is presented by Su et al. who performed GM on a strain of *Bacillus subtilis* (i.e., NCD-2), initially predicting its potential for the production of fengicin, surfactin, bacillaene, subtilosin, bacillibactin, bacillosin and other not previously reported molecules, that were later detected by UHPLC-QTOF-MS/MS in its fermentation extracts [62]. The increasing popularity of classical GM promoted the development of GM-specialized databases and novel bioinformatics tools with improved homology searching tools, specialized sequence analyses, and advanced prediction algorithms. A list of some currently available GM specialized databases and related bioinformatics tools are presented in **Tables 1** and **2**, respectively.

Currently, the most popular platform for GM of bacterial and fungal genomes is antiSMASH. It is up to now the most comprehensive by integrating its own database and incorporating different prediction tools [63]. The key of its popularity results from the integration of different complex secondary metabolite-specific gene analysis methods using a much more researcher-friendly interface [82]. Unfortunately, as shown in the tables, most advances have been made in bacteria and there is still a need to improve or create new bioinformatics tools to enable GM in other organisms such as fungi and especially plants, which commonly do not have biosynthetic gene clusters but a separated, often compartmentalized (cell type specific) synthesis of secondary metabolites, including transport of intermediates between cell types and even organs [83, 84].

2.2 Comparative genome mining

Classical GM alone fails to identify BGCs in genomic regions that do not follow a classical modular gene topology, as described by Donadio et al. since 1991. The

Database	Description	Ref.
antiSMASH database	Comprehensive resource on BGCs for secondary metabolites identified in bacterial genomes.	[63]
BACTIBASE	Open-access database used for the characterization of bacterial antimicrobial peptides.	[64]
ClusterMine360	Contains over 200 curated entries of BGCs clusters including classification of the potential compounds produced, taxonomic information of the producing organisms, and links to original data.	[65]
CSDB/r-CSDB	Manually curated database containing more than 160 PKS, NRPS, and PKS/NRPS BGCs.	[66]
DoBISCUIT	Contains a literature-based collection of BGCs for PKS and NRPS.	[67]
IMG-ABC	Contains automatically identified gene clusters, clusters with known biosynthesis products, and secondary metabolites.	[68]

Table 1.
Main databases focused on biosynthetic gene clusters (BGCs) encoding secondary metabolites.

Tool	Description	Ref.
antiSMASH	Fully automated tool for extracting genome data from bacteria and fungi to search for BGCs.	[69]
BiG-SCAPE	Uses the distance between BGCs (identified with antiSMASH), to create sequence similarity networks.	[70]
CLUSEAN	Allows homology searches and identification of conserved domains in BGCs of genes encoding for PKS and NRPS. Also classifies enzymes and predicts the domains specificity.	[71]
CLUSTER FINDER	Uses a probability approach to recognize BGCs in genomic and metagenomic data.	[72]
EvoMining	Uses phylogenetics to recognize, compare and identify BGCs associated with primary metabolism but that present a divergent phylogeny.	[73]
FunGeneClusterS	Allows the prediction of BGCs based on genomic and transcriptomic data for fungi.	[74]
MIPS-CG	Allows the identification of totally new BGCs using only genomic data.	[75]
NaPDoS	Detects and analyze genes associated with secondary metabolites.	[76]
PhytoClust	Detects BGCs of secondary metabolites in plant genomes.	[77]
PKMiner	Predicts novel BGCs of type II PKS and aromatic polyketide chemotypes using their conserved aromatase and cyclase domains.	[78]
plantiSMASH	An antiSMASH' version that uses plant genomes.	[79]
SBSPKS	Allows chemical analysis of experimentally characterized BGCs for PKS/NRPS proteins.	[80]
SMURF	Used for mining BGCs in fungi to identify conserved domains in PKS, NRPS, PKS/NRPS hybrids, and terpenoid genes.	[81]

Table 2.
Main tools for mining secondary metabolite biosynthesis gene clusters.

organization of open reading frames (ORFs) associated with secondary metabolite-producing genes that generally follow an order of distribution between catalytic and structural domains for modular PKSs or NRPSs, for example, is called a modular pattern [39]. These extensively described and annotated modules serve as a template for comparison with new sequences from available genomes [42].

Leblond and coworkers found more than 3300 BGCs for about 16,500 possible NRPS-associated enzymes in *Streptomyces ambofaciens*. However, when evaluating the potential enzymes *in silico*, they realized that many did not follow the modular pattern used as a template [85]. This, indeed, reduced the possibilities of modeling the possible secondary metabolites that could be produced by this bacterium. This is certainly an example of the current limitations of classical GM, which must contemplate new technologies (e.g., artificial intelligence (AI) and machine learning (ML)) in response to unconventional sequences that do not completely follow the expected organization.

One way to address these limitations is by integrating already existing tools that are focused more on the identification of patterns related to phylogeny and evolution instead of molecular function. For example, descriptions of lineage relationships can be made and some non-modular combinations of putative BGCs can be described between organisms that may not belong to the same taxonomic level. These results are not only valuable for the search for pathways to new natural products, but they also allow evolutionary reconstruction in the creation of metabolic pathways that respond to defense, competition, and attack of organisms in their ecosystem [86]. In plant metabolomics, such phylogentic relationships based on an untargeted fingerprint approach of natural products of different species were for the first time described in 2013 for *Urtica* species [87], still awaiting a full correlation with genomic data.

Two different ways of using phylogenetics approaches for comparative GM can be defined: In the first one, phylogenetics trees are constructed using both the whole sequences of the organisms under study and a pool of conserved well-characterized gene clusters associated to the production of a defined compound. In this way, BGC lineages can be traced and evolutionary relationships between apparently unrelated organisms can be established. Abdelmohsen et al. used this strategy to investigate biosynthetic pathways in actinomycetes isolated from marine sponges from the Red Sea. After a combination of taxonomic evaluation using the 16S ribosomal gene, PCR amplification of genes associated with modular PKS and NRPS, and phylogenetic analysis, the authors found that 20 of the actinomycetes isolates (speeded over 10 genera) possessed at least one of the biosynthetic genes analyzed [88]. This method has been extensively applied to identify novel potential BGCs [70, 89] and to create new gene clusters that can be further related to already annotated genomes of organisms previously studied at the experimental level.

The use of comparative GM has also allowed the identification of genes involved in the production of secondary metabolites in bacteria, by considering horizontal gene transfer events and phylogenetic analysis. Here, relationship trees are constructed using genes that are directly associated with the creation of specific compounds/secondary metabolites [90]. In this model, gene relationships are inferred primarily using the biosynthetic gene sequences only, and later those relationships are contrasted or strengthened by evaluating the rest of the organism's genome [91]. An example of the use of this method are studies conducted on the genus *Streptomyces*, where the production of secondary metabolites was again evaluated considering events of lateral gene transfer. It was found that, although horizontal gene transfer of the studied BGCs is not so frequent, the transfer of exogenous regulatory, resistance, and secondary metabolite production genes can significantly contribute to recombination events in those BGCs. Thus, comparative GM brings new relevant concepts such as the variable nature of those BGCs and their diversification even within very specific levels of phylogenetic discrimination. This undoubtedly paves the way not only to understand the evolution of BGCs in microorganisms but also to understanding the ecological landscape that it influences [91].

Currently, one of the methods to specifically evaluate putative catalytic domains in enzymes, using phylogenetic algorithms, is the Natural Product Domain Seeker (NaPDos), which organizes sequences into clades and allows the recognition of lineages of organisms capable of producing selected metabolites [76, 92]. This represents a new approach for the evaluation of possible non-homologous and undescribed enzymes (shown for modular PKS and NPRS) and to elucidate new chemical structures not yet identified. NaPDos initially contained only data from PCR fragments but now is a comprehensive tool that also includes genomics and metagenomics data [93]. This is particularly important because it allows the evaluation of genomic data obtained from complex samples such as soils, sediments, water sources, wastes, etc. (metagenomics). With NaPDos it is even possible to estimate the diversity of microorganisms from the sampled source, as well as to evaluate the genetic potential for the biosynthesis of different metabolites [93].

2.3 Genome mining in synthetic biology

The identification of novel BGCs resulting from genomic mining studies represents a great opportunity for synthetic biologists and metabolic engineering as it allows the identification, construction, synthesis, and expression of BioBricks in heterologous models or to discover natural compounds with outstanding properties. One of the most significant commercial examples of this application has been observed during the engineering of yeast for the biosynthesis of valuable products such as artemisinin (an antimalarial drug) by using BioBricks identified through GM [35, 94]. Recently, GM has been also used to identify more than 70 syntheses involved in the production of hypermodified peptide cytotoxins (i.e., unique, and valuable chemotherapeutics) by mining prokaryotic diversity [95]. With the help of GM, the identification of several cryptic metabolic pathways has been possible, giving way to combinatorial biosynthesis, which can be used in the construction of biosynthetic units, following the pattern of BGCs. These approaches also present challenges mainly related to our current understanding of the interdependent metabolic circuits, and the complexity in tracking them. This will certainly require many more efforts from bioinformatics to enrich genomic mining by including additional omics data such as transcriptomics, metabolomics, and proteomics not only for microorganisms but also for eukaryotes with their complexer, usually unclustered biosynthetic production networks [96].

2.4 Transcriptome mining

A transcriptome represents a "snapshot" of a RNA population in a certain tissue or at a specific developmental stage. Compared to the genomic information of the same organism, a transcriptomic dataset is less complex as it does not contain any information, for example, on the untranslated regions of a genome (e.g., promoters). Transcriptomes also do not provide information on the physical organization of the individual genetic elements—a fact which in turn represents an obstacle for the application of classical GM methods (see previous sections) used, for instance, for pathway elucidation in plants. However, several advantages make transcriptome mining (TM) a valuable alternative in the last years: First, unlike in a "static" genome, differential analysis is possible for transcriptomic data. Thus, the identification of tissue-specific transcripts (pathways restricted to special organs) and discrimination of non-functional RNAs (pseudogenes) is much easier than in GM approaches. Secondly, the less complex datasets facilitate mining in organisms with large and complex genomes such as plants [97], which in general developed multi-member gene families with redundant functions during evolution. In conjunction with the fact that the organization of biosynthetic pathways into gene clusters is exceptional in plants [98], TM is increasingly used in this class of organisms to mine for NP pathways as well as to study different aspects of plant physiology. Recent examples for the latter purpose include the dissection of the response to changing temperatures [99], drought stress [100], or defense against pathogens in model and non-model plants [101, 102].

First reports on TM used for the discovery of NP biosynthetic genes date back to the first decade of the 21st century. The reports were based on so-called expressed sequence tag (EST) databases [103], which were developed as an alternative to earlier microarray-driven methods for expression analysis. Milestones for the application in the plant field were the establishment of specific EST databases [104] and the access to programs that used both microarray data and transcriptome datasets in the frame of transcriptome profiling (e.g., eVOC [105]). Continued software development led to more advanced approaches which integrated data modeling in targeted plant engineering [106]. Alongside with the use of co-expression analysis as a standard tool in multifaceted mining strategies [107] and the current decrease in prices for transcriptome sequencing, the developments led to a continuous increase in the annual output of TM-based publications (3 in 2003, 84 in 2020).

For instance, all classes of NPs found in plants were targeted using TM in the last years. Most reports focused on **terpenoids**, including papers on the identification of single enzymes such as terpene cyclases/synthases [108], associated biocatalysts [109] or comparative evolutionary studies of genes in whole plant families such as Pinaceae [110] or Lamiaceae [111]. An outstanding example is the mining for biocatalysts involved in the biosynthesis of the insecticidal limonoid azadirachtin in neem (*Azadirachta indica*) [112]. By using a comparative analysis of three limonoid-containing species from the order Sapindales, the authors could identify key enzymes involved in the early steps of the pathway, namely the initial terpene cyclase forming the basal triterpene scaffold and subsequent cytochromes involved in tailoring modifications. In the field of **alkaloids**, TM was similarly applied, yielding the enzyme norbelladine synthase from *Narcissus pseudonarcissus* [113]. This enzyme, which is used for a coupling step during the synthesis of the anticancer agent galantamine in *Narcissus* species, was fished by a TM-based screening for functional homologs of an enzyme catalyzing a similar enzymatic reaction in opium poppy. Hagel and co-workers [114] used a similar but broader approach to compare plants with a pronounced production of benzylisoquinoline alkaloids. Differential analysis of the transcriptomes and metabolomes of 20 species from the order Ranunculales revealed 850 genes that are potentially involved in alkaloid biosynthesis and are interesting candidates for use in alkaloid Synthetic Biology. A noteworthy example concerning the biosynthesis of plant **phenolics** is the study of Lau and Satteley [115], which describes mining for enzymes required for the production of podophyllotoxin. This lignan is an antiviral polyphenol isolated from mayapple (*Podophyllum peltatum*), and six of the enzymes involved in its biosynthesis could be identified by TM followed by subsequent co-expression in tobacco. Another example is the insight from TM and Metabolomics in the synthesis of hypericin in the medicinal plant St. John's wort (*Hypericum perforatum*) [116].

Future studies will certainly use extensive TM to further explore the biosynthetic machineries to high-value metabolites other than terpenes, alkaloids, and phenolics. In agreement with this assumption, the latest reports on TM already target pathways to antimicrobial cyclopeptides [117], polysaccharides [118], or compounds derived from fatty acids [119]. In general, TM studies will definitely benefit from the integration of multi-level omics data in the future. Such comprehensive methods have already been applied in proof-of-concept studies, including the combination of TM with proteomics to mine for cyclopeptides [120] or in-plant "regulomics", i.e., in software tools comparing transcriptomes with (epi)genomic data to identify regulatory networks [121].

3. Metabolic data mining

Metabolism is typically defined as the sum of pathways and cycles representing all the sets of biochemical reactions occurring at a cell and in which the product of a particular chemical reaction becomes the substrate of the subsequent reaction [122]. Certainly, the understanding of this concept is key in the realm of biological sciences, especially in the post-genomic era, where we have embraced a paradigm shift from a gene-centered view to an increasing interest in omics-driven high-throughput data types, sources, and approaches [123]. In line with the current move towards systems biology, the mining of metabolism data (metabolic data mining) includes not only the systematic study of component metabolites (i.e., **metabolomics**) [124], but also of all the controlled biochemical reactions in an organism responsible for their production, which is more recently understood under the name of **reactomics** [125] and related processes such as in **fluxomics**

[126, 127]. In metabolomics, numerous subclasses have emerged, as in distinction to especially genomics, a really holistic determination of the metabolome is impossible: no method exists to extract and analyze all metabolites of an organism completely in one experiment. Unlike in genomics, transcriptomics or proteomics, metabolome analytics cannot rely on a one dimensional sequential biopolymer of a limited number of monomer units and a few handful of derivatizations (methylation, post-translational modifications etc.). Instead, most compounds are unique, they are rarely produced by linear monomer assembly processes which can be deconvoluted by standardized processes. But instead a metabolome is a mixture of compounds with highly complex 2D and mostly 3D molecular structures of maximum variability and physicochemical property divergenceies (e.g., sugars vs. triglycerides). Subclasses have thus emerged, e.g., lipidomics or glycomics. Along with the great advances of computing technologies, all types of studies -especially when applied in combination- have led us to witness an unprecedented revolution in biotechnology by finding patterns or trends that explain the behavior of large data sets in a specific context and as automated as possible. Thus, during the last decade, a large number of metabolic pathways have been mined to identify the key elements and modules for the production of drugs, foods, fuels, and a plethora of bioactive compounds [128–130], including the combination of transcriptome and metabolome studies [116].

The trifold correlation of metabolomic, transcriptomic/genomic and phenotypical data ideally allows to identify both gene loci responsible and the biosynthetic components responsible for a property (phenotype), the biosynthetic pathways for their production, and the genetic control elements associated with them (GWAS—genome wide association study). This allows e.g., improved molecular breeding in plants without the necessity of producing GMOs. An example is a study on downy mildew resistance in hops (*Humulus lupulus*), i.e., tackling it most devastating pathogen by identifying the intrinsic strengths of its chemical defense. The identification of key metabolites responsible for mildew resistance, their associated pathways and genetic breeding markers associated with downy mildew resistance now allows the targeted (non-GMO) molecular breeding of resistant phenotypes [131]. The same tools can, of course, also be used for higher production using genetic improvement (GMOs) [80]. The different strategies for the identification of these metabolic pathways via data collection and coupling, reactome reconstruction, and rational exploration of the chemical space will be further discussed.

3.1 Metabolic data collection and coupling

A typical workflow in metabolic data mining aimed to elucidate interaction networks and reactomes is shown in **Figure 1**. Initially, metabolic data is collected including information on enzymes and metabolites. Then, the recognition and

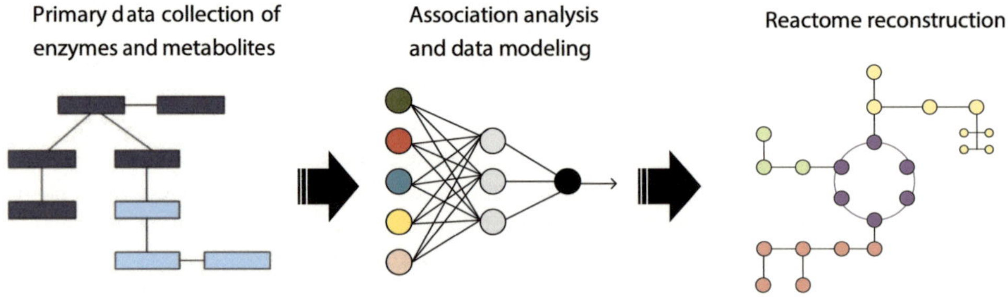

Figure 1.
Standard workflow in metabolic data mining to elucidate interaction networks and reactomes.

coupling of network patterns are carried out by association analysis and data modeling to obtain a reduction in data dimensionality. Finally, reactomes are reconstructed to elucidate the corresponding network dynamics and topology [132]. This knowledge forms the basis for future metabolic engineering experiments aimed to enhance the production of the desired compound or to assemble novel native but also synthetic/unnatural biosynthetic pathways. Interestingly, the current advances in the development of novel BioBricks and the design of novel artificial metabolic networks promote the rapid and efficient coupling of a series of biological parts into a highly reusable large-scale framework [133].

3.2 Proficient exploration of chemical space: natural products and fragments

Metabolic data mining also may involve the use of small compounds derived from the primary and, most especially, secondary metabolism of living organisms. These metabolites, typically referred to as natural products (NPs), have largely been used as a source of chemical entities with promising physicochemical, medicinal or other features, being used directly (unmodified), as a substructure, or as inspiration for a structurally similar chemical scaffold [134, 135]. NPs have been used for ages as medicines than the synthetic bioactives and as scaffolds for the rational design of novel synthetic drugs [136, 137]. Interestingly, they occupy a much larger fraction of the ensemble of all chemical compounds (i.e., have a larger structural diversity), which is classically known among theoretical and computational chemists as **chemical space** ($\sim 10^{60}$ molecules) [138, 139]. In the field of medicinal chemistry, and considering we only know just a bit portion of the estimated chemical space ($\sim 10^{8}$ molecules) [140], the use of NP-based libraries represents a priceless opportunity for scientists to make bigger and faster leaps within it [141, 142]. This fact represents an additional advantage taking into account that conventional combinatorial chemistry (usually termed combichem) without input from natural products initially had very limited success in novel drug discovery [141, 143], having its strength rather in optimization in most cases [141]. On the other hand, an alternative scenario intended to explore the chemical space more profoundly and, thus, may be used to harness metabolic data involves the principles of molecular fragmentation. According to this technique, a chemical compound of interest is not identified and evaluated as a whole, but instead, it is developed starting from structural molecular components usually within the range 120–300 Da (i.e., fragments) [144, 145]. Although many current chemical libraries are available as fragments per se, various cleavage methods such as RECAP (Retrosynthetic Combinatorial Analysis Procedure) have been widely used to deconstruct chemical libraries of both NPs and other classes of chemical entities [146, 147]. Among the many advantages of using fragments are not only their potential to navigate into the chemical space in a more cost-effective manner compared, for example, to drug-sized molecules, but also their potential to favor the protein-ligand complementarity and facilitate selectivity adjustments during optimization processes (a more detailed description is given in **Figure 2**) [148, 149]. Once more, within the field of BioBricks, the possibility of understanding every fragment as an independent brick could facilitate not only the recovery of specific substructures during a virtual screening (VS) protocol but also the coupling of the best combinations of substructures to obtain a final candidate for further development. It is worth mentioning that fragments could be “recycled” to be considered in the development of a bigger compound if other partner fragments can supply -and balance- particular physicochemical properties of interest. This is fully illustrated in terms of ligand efficiency (LE) metrics as a phenomenon called fragment “rescue” effect [150]. Through an application of these kinds of concepts and approaches, the

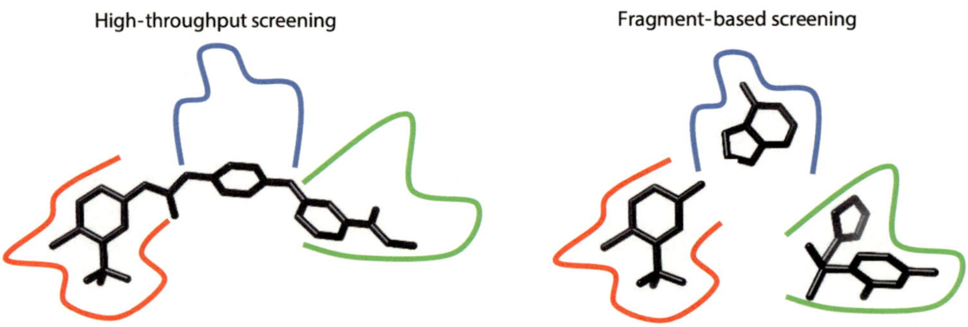

Figure 2.
Comparison between typical high-throughput screening and fragment-based screening. *In the left panel, it is evident that although one specific part of the drug compound exhibits a good fit within most of the pocket of a hypothetical target protein (red curved line), the other two parts of the same compound do not occupy any specific binding (blue curved line) or occupies subsites of the active center only partially (green curved line). In contrast, the right panel shows that the consideration of fragments for screening allowed the identification of chemical entities with high inherent affinity to the corresponding pockets. Although only shape and size are included in the illustration for clarity, many other physicochemical characteristics such as lipophilicity and charge may affect the complementarity between a chemical moiety and its target receptor.*

scientific community may benefit from metabolomic data mining of compounds able to mediate diverse functions in biological systems.

4. Conclusions

Multi-omics data mining has revolutionized science by enabling overlaps among different fields of study such as biochemistry, molecular biology, synthetic biology, organic and medicinal chemistry, computational chemistry, chemical engineering, and high-performance computing. This represents a crucial breakthrough that is expected to accelerate our comprehension of complex biological systems and, most interestingly, the identification, selection, and recovery of novel pieces of biological information in the form of BioBricks for the design of biofactories. Currently, we have unprecedented access to large multi-omics data repositories, which make possible the discovery, identification, and coupling of these BioBricks. This is an important step to unleash different biological functions, or to rationally design metabolic pathways for the biosynthesis of valuable products. However, there is still a need for integrating additional cutting-edge technologies in computing and data science such as machine learning, artificial intelligence, and big and smart data analytics that can further boost the discovery and *de novo* design of BioBricks with high impact in pharma, cosmetics, fine chemical and nutraceutical industries.

Conflict of interest

The authors declare no conflict of interest.

Author details

Angie Burgos-Toro[1], Martin Dippe[2], Andres Felipe Vásquez[3,4], Eric Pierschel[2], Ludger Aloisius Wessjohann[2] and Miguel Fernández-Niño[2]*

1 Department of Biology, Universidad Nacional de Colombia, Colombia

2 Department of Bioorganic Chemistry, Leibniz-Institute of Plant Biochemistry, Halle, Germany

3 Grupo de Diseño de Productos y procesos (GDPP), Facultad de Ingeniería, Universidad de los Andes, Bogotá, Colombia

4 Naturalius S.A.S., Bogotá, Colombia

*Address all correspondence to: mfernand@ipb-halle.de

References

[1] Hood L, Rowen L. The human genome project: Big science transforms biology and medicine. Genome Medicine. 2013;**5**:1-8. DOI: 10.1186/GM483

[2] Watson JD, Crick FHC. Molecular structure of nucleic acids: A structure for deoxyribose nucleic acid. Nature. 1953;**171**:737-738. DOI: 10.1038/171737a0

[3] Sanger F, Nicklen S, Coulson A. DNA sequencing with chain-terminating inhibitors. Proceedings of the National Academy of Sciences of the United States of America. 1977;**74**:5463-5467. DOI: 10.1073/PNAS.74.12.5463

[4] Chidgeavadze ZG, Beabealashvilli RS, Atrazhev AM, Kukhanova MK, Azhayev AV, Krayevsky AA. 2′,3′-Dideoxy-3′ amlnonudeo 5′ triphosphates are the terminators of DNA synthesis catalyzed by DNA polymerases. Nucleic Acids Research. 1984;**12**:1671-1686. DOI: 10.1093/NAR/12.3.1671

[5] Smith LM, Fung S, Hunkapiller MW, Hunkapiller TJ, Hood LE. The synthesis of oligonucleotides containing an aliphatic amino group at the 5′ terminus: Synthesis of fluorescent DNA primers for use in DNA sequence analysis. Nucleic Acids Research. 1985;**13**:2399-2412. DOI: 10.1093/NAR/13.7.2399

[6] Ansorge W, Sproat BS, Stegemann J, Schwager C. A non-radioactive automated method for DNA sequence determination. Journal of Biochemical and Biophysical Methods. 1986;**13**:315-323. DOI: 10.1016/0165-022X(86)90038-2

[7] Ansorge W, Sproat B, Stegemann J, Schwager C, Zenke M. Automated DNA sequencing: Ultrasensitive detection of fluorescent bands during electrophoresis. Nucleic Acids Research. 1987;**15**:4593-4602. DOI: 10.1093/NAR/15.11.4593

[8] Prober JM, Trainor GL, Dam RJ, Hobbs FW, Robertson CW, Zagursky RJ, Cocuzza AJ, Jensen MA, Baumeister K, et al. A system for rapid DNA sequencing with fluorescent chain-terminating dideoxynucleotides. Science. 1987;**238**:336-341. DOI: 10.1126/SCIENCE.2443975

[9] Kambara H, Nishikawa T, Katayama Y, Yamaguchi T. Optimization of parameters in a DNA sequenator using fluorescence detection. Nature Biotechnology. 1988;**6**:816-821. DOI: 10.1038/nbt0788-816

[10] Luckey JA, Drossman H, Kostichka AJ, Mead DA, D'Cunha J, Norris TB, et al. High speed DNA sequencing by capillary electrophoresis. Nucleic Acids Research. 1990;**18**:4417-4421. DOI: 10.1093/NAR/18.15.4417

[11] Swerdlow H, Gesteland R. Capillary gel electrophoresis for rapid, high resolution DNA sequencing. Nucleic Acids Research. 1990;**18**:1415-1419. DOI: 10.1093/NAR/18.6.1415

[12] Jackson DA, Symons RH, Berg P. Biochemical method for inserting new genetic information into DNA of simian virus 40: Circular SV40 DNA molecules containing lambda phage genes and the galactose operon of *Escherichia coli*. Proceedings of the National Academy of Sciences. 1972;**69**:2904-2909. DOI: 10.1073/PNAS.69.10.2904

[13] Cohen SN, Chang ACY, Boyer HW, Helling RB. Construction of biologically functional bacterial plasmids in vitro. Proceedings of the National Academy of Sciences. 1973;**70**:3240-3244. DOI: 10.1073/PNAS.70.11.3240

[14] Nyrén P, Lundin A. Enzymatic method for continuous monitoring of

inorganic pyrophosphate synthesis. Analytical Biochemistry. 1985;**151**: 504-509. DOI: 10.1016/0003-2697(85) 90211-8

[15] Voelkerding KV, Dames SA, Durtschi JD. Next-generation sequencing: From basic research to diagnostics. Clinical Chemistry. 2009; **55**:641-658. DOI: 10.1373/ CLINCHEM.2008.112789

[16] Bentley DR, Balasubramanian S, Swerdlow HP, Smith GP, Milton J, Brown CG, et al. Accurate whole human genome sequencing using reversible terminator chemistry. Nature. 2008;**456**:53-59. DOI: 10.1038/ nature07517

[17] Fedurco M, Romieu A, Williams S, Lawrence I, Turcatti G. BTA, a novel reagent for DNA attachment on glass and efficient generation of solid-phase amplified DNA colonies. Nucleic Acids Research. 2006;**34**:e22-e22. DOI: 10.1093/NAR/GNJ023

[18] McKernan KJ, Peckham HE, Costa GL, McLaughlin SF, Fu Y, Tsung EF, et al. Sequence and structural variation in a human genome uncovered by short-read, massively parallel ligation sequencing using two-base encoding. Genome Research. 2009;**19**:1527-1541. DOI: 10.1101/GR.091868.109

[19] Drmanac R, Sparks AB, Callow MJ, Halpern AL, Burns NL, Kermani BG, et al. Human genome sequencing using unchained base reads on self-assembling DNA nanoarrays. Science. 2010;**327**: 78-81. DOI: 10.1126/SCIENCE.1181498

[20] Rothberg JM, Hinz W, Rearick TM, Schultz J, Mileski W, Davey M, et al. An integrated semiconductor device enabling non-optical genome sequencing. Nature. 2011;**475**:348-352. DOI: 10.1038/nature10242

[21] Braslavsky I, Hebert B, Kartalov E, Quake SR. Sequence information can be obtained from single DNA molecules. Proceedings of the National Academy of Sciences. 2003;**100**:3960-3964. DOI: 10.1073/PNAS.0230489100

[22] Harris TD, Buzby PR, Babcock H, Beer E, Bowers J, Braslavsky I, et al. Single-molecule DNA sequencing of a viral genome. Science. 2008;**320**:106-109. DOI: 10.1126/SCIENCE.1150427

[23] Thermes C. Ten years of next-generation sequencing technology. Trends in Genetics. 2014;**30**:418-426. DOI: 10.1016/J.TIG.2014.07.001

[24] Levene MJ, Korlach J, Turner SW, Foquet M, Craighead HG, Webb WW. Zero-mode waveguides for single-molecule analysis at high concentrations. Science. 2003;**299**:682-686. DOI: 10.1126/SCIENCE.1079700

[25] Haque F, Li J, Wu HC, Liang XJ, Guo P. Solid-state and biological nanopore for real-time sensing of single chemical and sequencing of DNA. Nano Today. 2013;**8**:56-74. DOI: 10.1016/J. NANTOD.2012.12.008

[26] Kasianowicz JJ, Brandin E, Branton D, Deamer DW. Characterization of individual polynucleotide molecules using a membranechannel. Proceedings of the National Academy of Sciences. 1996;**93**: 13770-13773. DOI: 10.1073/ PNAS.93.24.13770

[27] Clarke J, Wu H-C, Jayasinghe L, Patel A, Reid S, Bayley H. Continuous base identification for single-molecule nanopore DNA sequencing. Nature Nanotechnology. 2009;**4**:265-270. DOI: 10.1038/nnano.2009.12

[28] Eisenstein M. Oxford nanopore announcement sets sequencing sector abuzz. Nature Biotechnology. 2012;**30**: 295-296. DOI: 10.1038/NBT0412-295

[29] Loman NJ, Quinlan AR. Poretools: A toolkit for analyzing nanopore sequence

data. Bioinformatics. 2014;**30**: 3399-3401. DOI: 10.1093/ BIOINFORMATICS/BTU555

[30] Branton D, Deamer DW, Marziali A, Bayley H, Benner SA, Butler T, et al. The potential and challenges of nanopore sequencing. Nature Biotechnology. 2008;**26**: 1146-1153. DOI: 10.1038/nbt.1495

[31] Messana I, Cabras T, Iavarone F, Vincenzoni F, Urbani A, Castagnola M. Unraveling the different proteomic platforms. Journal of Separation Science. 2013;**36**:128-139. DOI: 10.1002/ JSSC.201200830

[32] Rochfort S. Metabolomics reviewed: A new "Omics" platform technology for systems biology and implications for natural products research. Journal of Natural Products. 2005;**68**:1813-1820. DOI: 10.1021/NP050255W

[33] Foulston L. Genome mining and prospects for antibiotic discovery. Current Opinion in Microbiology. 2019; **51**:1-8. DOI: 10.1016/j.mib.2019.01.001

[34] Zerikly M, Challis GL. Strategies for the discovery of new natural products by genome mining. Chembiochem. 2009;**10**:625-633. DOI: 10.1002/ cbic.200800389

[35] Albarano L, Esposito R, Ruocco N, Costantini M. Genome mining as new challenge in natural products discovery. Marine Drugs. 2020;**18**:1-17

[36] Martin JF, Liras P. Organization and expression of genes involved in the biosynthesis of antibiotics and other secondary metabolites. Annual Review of Microbiology. 1989;**43**:173-206. DOI: 10.1146/annurev. mi.43.100189.001133

[37] Wright L, Hopwood D. Identification of the antibiotic determined by the SCPl. Journal of General Microbiology. 1975;**95**:96-106

[38] Ziemert N, Alanjary M, Weber T. Natural product reports the evolution of genome mining in microbes—A review. Natural Product Reports. 2016;**33**: 988-1005. DOI: 10.1039/C6NP00025H

[39] Donadio S, Staver MJ, Mcalpine JB, Swanson SJ, Katz L. Modular organization of genes required for complex polyketide biosynthesis. Science. 1991;**205**:675-679

[40] Beutler B, Hoebe K, Du X, Ulevitch RJ. How we detect microbes and respond to them: The Toll-like receptors and their transducers. Journal of Leukocyte Biology. 2003;**74**:479-485. DOI: 10.1189/jlb.0203082

[41] Bachmann BO, Van Lanen SG, Baltz RH. Microbial genome mining for accelerated natural products discovery: Is a renaissance in the making? Journal of Industrial Microbiology & Biotechnology. 2014;**41**:175-184. DOI: 10.1007/s10295-013-1389-9

[42] Nett M. Genome mining: Concept and strategies for natural product discovery. Progress in the Chemistry of Organic Natural Products. 2014;**99**:199-245. DOI: 10.1007/978-3-319-04900-7_4.

[43] Baltz RH. Synthetic biology, genome mining, and combinatorial biosynthesis of NRPS—Derived antibiotics: A perspective. Journal of Industrial Microbiology and Biotechnology. 2018;**45**:635-649. DOI: 10.1007/s10295-017-1999-8

[44] Sekurova ON, Schneider O, Zotchev SB. Novel bioactive natural products from bacteria via bioprospecting, genome mining and metabolic engineering. Microbial Biotechnology. 2019;**12**:828-844. DOI: 10.1111/1751-7915.13398

[45] Omura S, Ikeda H, Ishikawa J, Hanamoto A, Takahashi C, Shinose M, et al. Genome sequence of an industrial microorganism *Streptomyces avermitilis*:

Deducing the ability of producing secondary metabolites. PNAS. 2001;**98**: 12215-12220

[46] Chater KF. Genetics of differentiation in streptomyces. Annual Review of Microbiology. 1993;**47**:685-713

[47] Behnken S, Hertweck C. Anaerobic bacteria as producers of antibiotics. Applied Microbiology and Biotechnology. 2012;**96**:61-67. DOI: 10.1007/s00253-012-4285-8

[48] Welker M, Dittmann E, Von Döhren H. Cyanobacteria as a source of natural products. Methods in Enzymology. 2012;**517**:23-46. DOI: 10.1016/B978-0-12-404634-4.00002-4

[49] Katz M, Hover BM, Brady SF. Culture-independent discovery of natural products from soil metagenomes. Journal of Industrial Microbiology & Biotechnology. 2016; **43**:129-141. DOI: 10.1007/s10295-015-1706-6

[50] Zhang X, Wang TT, Xu QL, Xiong Y, Zhang L, Han H, et al. Genome mining and comparative biosynthesis of meroterpenoids from two phylogenetically distinct fungi. Angewandte Chemie. 2018;**180**: 8184–8188 DOI: 10.1002/ ange.201804317

[51] Medema MH, Cimermancic P, Sali A, Takano E, Fischbach MA. A systematic computational analysis of biosynthetic gene cluster evolution: Lessons for engineering biosynthesis. PLoS Computational Biology. 2014;**10**: e1004016. DOI: 10.1371/journal. pcbi.1004016

[52] Hadjithomas M, Chen IA, Chu K, Ratner A, Palaniappan K, Szeto E, et al. IMG-ABC: A knowledge base to fuel discovery of biosynthetic gene clusters and novel secondary metabolites. MBio. 2015;**6**:1-10. DOI: 10.1128/ mBio.00932-15.Editor

[53] Nogales A, Martínez-sobrido L. Reverse genetics approaches for the development of influenza vaccines. International Journal of Molecular Sciences. 2017;**18**:1-26. DOI: 10.3390/ ijms18010020

[54] Farnet CM, Zazopoulos E. Improving drug discovery from microorganisms. In: Natural Products: Drug Discovery and Therapeutic Medicine. 2005. pp. 95-106. Humana Press (Totowa, US). DOI: 10.1007/978-1-59259-976-9_5

[55] Timmermans ML, Paudel YP, Ross AC. Investigating the biosynthesis of natural products from marine proteobacteria: A survey of molecules and strategies. Marine Drugs. 2017;**15**. p. 235 DOI: 10.3390/md15080235

[56] Lee M, Philippe J, Katsanis N, Zhou W. Polyketide synthase plays a conserved role in otolith formation. Zebrafish. 2019;**16**:363-369. DOI: 10.1089/zeb.2019.1734

[57] Adhikari K, Lo I, Chen C, Wang Y, Lin K, Zadeh SM, et al. Chemoenzymatic synthesis and biological evaluation for bioactive molecules derived from bacterial benzoyl coenzyme A ligase and plant type III polyketide synthase. Biomolecules. 2020;**10**. p. 738

[58] Maansson M, Vynne NG, Klitgaard A, Nybo JL, Melchiorsen J, Nguyen DD, et al. An integrated metabolomic and genomic mining workflow to uncover the biosynthetic potential of bacteria. mSystems. 2016;**1**: 1-14. DOI: 10.1128/mSystems.00028-15. Editor

[59] Olano C, Méndez C, Salas JA. Strategies for the design and discovery of novel antibiotics using genetic engineering and genome mining. In: Antimicrobial Compounds. 2014. pp. 1-25. Springer (Berlin, Germany). DOI: 10.1007/978-3- 87 642-40444-3

[60] Hannigan GD, Prihoda D, Palicka A, Soukup J, Klempir O, Rampula L, et al. A deep learning genome-mining strategy for biosynthetic gene cluster prediction. Nucleic Acids Research. 2019;***47e110***. DOI: 10.1093/nar/gkz654

[61] Russell AH, Truman AW. Genome mining strategies for ribosomally synthesised and post-translationally modified peptides. Computational and Structural Biotechnology Journal. 2020; **18**:1838-1851. DOI: 10.1016/j.csbj.2020.06.032

[62] Su Z, Chen X, Liu X, Guo Q, Li S, Lu X, et al. Genome mining and UHPLC-QTOF-MS/MS to identify the potential antimicrobial compounds and determine the specificity of biosynthetic gene clusters in Bacillus subtilis NCD-2. BMC Genomics. 2020;**21**:1-16

[63] Blin K, Shaw S, Steinke K, Villebro R, Ziemert N, Lee Y, et al. antiSMASH 5. 0: Updates to the secondary metabolite genome mining pipeline. Nucleic Acids Research. 2019; **47**:81-87. DOI: 10.1093/nar/gkz310

[64] Hammami R, Zouhir A, Le Lay C, Ben HJ, Fliss I. BACTIBASE second release: A database and tool platform for bacteriocin characterization. BMC Microbiology. 2010;**10**:1-5

[65] Tremblay N, Hill P, Conway KR, Boddy CN. The use of ClusterMine360 for the analysis of polyketide and nonribosomal peptide biosynthetic pathways. Methods in Molecular Biology. 2016;1401:233-52. DOI: 10.1007/978-1-4939-3375-4_15

[66] Diminic J, Zucko J, Trninic I, Cullum J, Starcevic A. Databases of the thiotemplate modular systems (CSDB) and their in silico recombinants (r-CSDB). Journal of Industrial Microbiology and Biotechnology. 2013; **40**:653-659. DOI: 10.1007/s10295-013-1252-z

[67] Ichikawa N, Sasagawa M, Yamamoto M, Komaki H, Yoshida Y. DoBISCUIT: A database of secondary metabolite biosynthetic gene clusters. Nucleic Acids Research. 2013;**41**:408-414. DOI: 10.1093/nar/gks1177

[68] Palaniappan K, Chen IA, Chu K, Ratner A, Seshadri R, Kyrpides NC, et al. IMG-ABC v.5.0: An update to the IMG/atlas of biosynthetic gene clusters knowledgebase. Nucleic Acids Research. 2019;**48**:422-430. DOI: 10.1093/nar/gkz932

[69] Alanjary M, Blin K, Weber T, Medema MH, Ziemert N. ARTS 2.0: Feature updates and expansion of the Antibiotic Resistant Target Seeker for comparative genome mining. Nucleic Acids Research. 2020;**48**:546-552. DOI: 10.1093/nar/gkaa374

[70] Navarro-Muñoz JC, Selem-mojica N, Mullowney MW, Kautsar SA, Tryon JH, Parkinson EI, et al. A computational framework to explore large-scale biosynthetic diversity. Nature Chemical Biology. 2020;**16**:60-68. DOI: 10.1038/s41589-019-0400-9

[71] Weber T, Rausch C, Lopez P, Hoof I, Gaykova V, Huson DH, et al. CLUSEAN: A computer-based framework for the automated analysis of bacterial secondary metabolite biosynthetic gene clusters. Journal of Biotechnology. 2009;**140**:13-17. DOI: 10.1016/j.jbiotec.2009.01.007

[72] Cimermancic P, Medema MH, Claesen J, Kurita K, Brown LCW, Mavrommatis K, et al. Insights into secondary metabolism from a global analysis of prokaryotic biosynthetic gene clusters. Cell. 2010;**158**:412-421. DOI: 10.1016/j.cell.2014.06.034

[73] Cruz-Morales P, Kopp JF, Martı C, Barona-go F, Selem-mojica N, Ramos-aboites H. Phylogenomic analysis of natural products biosynthetic gene

clusters allows discovery of arseno-organic metabolites in model streptomycetes. Genome Biology and Evolution. 2016;**8**:1906-1916. DOI: 10.1093/gbe/evw125

[74] Andersen MR, Nielsen JB, Klitgaard A, Petersen LM, Zachariasen M, Hansen TJ. Accurate prediction of secondary metabolite gene clusters in filamentous fungi. PNAS. 2012;**110**: 100-107. DOI: 10.1073/pnas.1205532110

[75] Umemura M, Koike H, Machida M. Motif-independent de novo detection of secondary metabolite gene clusters—Toward identification from filamentous fungi. Frontiers in Microbiology. 2015;**6**: 1-14. DOI: 10.3389/fmicb.2015.00371

[76] Ziemert N, Podell S, Penn K, Badger JH, Allen E, Jensen PR. The natural product domain seeker NaPDoS: A phylogeny based bioinformatic tool to classify secondary metabolite gene diversity. PLoS One. 2012;**7**:1-9. DOI: 10.1371/journal.pone.0034064

[77] Nadine T, Fuchs L, Aharoni A. The PhytoClust tool for metabolic gene clusters discovery in plant genomes. Nucleic Acids Research. 2017;**45**: 7049-7063. DOI: 10.1093/nar/gkx404

[78] Kim J, Yi G. PKMiner: A database for exploring type II polyketide synthases. BMC Microbiology. 2012;**12**: 1-12

[79] Blin K, Andreu P, Santos ELCDL, Del Carratore F, Lee SY, Medema MH, et al. The antiSMASH database version 2: A comprehensive resource on secondary metabolite biosynthetic gene. Nucleic Acids Research. 2019;**47**: 625-630. DOI: 10.1093/nar/gky1060

[80] Staniek A, Bouwmeester H, Fraser PD, Kayser O, Martens S, Tissier A, et al. Natural products—Modifying metabolite pathways in plants. Biotechnology Journal. 2013;**8**:1159-1171. DOI: 10.1002/BIOT.201300224

[81] Staniek A, Bouwmeester H, Fraser PD, Kayser O, Martens S, Tissier A, et al. Natural products—Learning chemistry from plants. Biotechnology Journal. 2014;**9**:326-336. DOI: 10.1002/BIOT.201300059

[82] Kautsar SA, Duran HGS, Blin K, Osbourn A, Medema H. plantiSMASH: Automated identification, annotation and expression analysis of plant biosynthetic gene clusters. Nucleic Acids Research. 2017;**45**:55-63. DOI: 10.1093/nar/gkx305

[83] Khater S, Gupta M, Agrawal P, Sain N, Prava J, Gupta P, et al. SBSPKSv2: Structure-based sequence analysis of polyketide synthases and non-ribosomal peptide synthetases. Nucleic Acids Research. 2017;**45**:72-79. DOI: 10.1093/nar/gkx344

[84] Khaldi N, Seifuddin FT, Turner G, Haft D, Nierman WC, Wolfe KH, et al. SMURF: Genomic mapping of fungal secondary metabolite clusters. Fungal Genetics and Biology. 2010;**47**: 736-741. DOI: 10.1016/j.fgb.2010.06.003

[85] Leblond P, Gondry M, Juguet M, Lautru S. An iterative nonribosomal peptide synthetase assembles the pyrrole-amide antibiotic congocidine in *Streptomyces ambofaciens*. Cell Chemical Biology. 2009;**2820**:421-431. DOI: 10.1016/j.chembiol.2009.03.010

[86] Kang HS. Phylogeny—Guided (meta) genome mining approach for the targeted discovery of new microbial natural products. Journal of Industrial Microbiology & Biotechnology. 2017; **44**:285-293. DOI: 10.1007/s10295-016-1874-z

[87] Farag MA, Weigend M, Luebert F, Brokamp G, Wessjohann LA. Phytochemical, phylogenetic, and anti-inflammatory evaluation of 43 Urtica accessions (stinging nettle) based on UPLC-Q-TOF-MS metabolomic

profiles. Phytochemistry. 2013;**96**: 170-183. DOI: 10.1016/J. PHYTOCHEM.2013.09.016

[88] Abdelmohsen UR, Yang C, Horn H, Hajjar D, Ravasi T, Hentschel U. Actinomycetes from red sea sponges: Sources for chemical and phylogenetic diversity. Marine Drugs. 2014;**12**: 2771-2789. DOI: 10.3390/md12052771

[89] Singh SP, Klisch M, Sinha RP, Häder D. Genome mining of mycosporine-like amino acid (MAA) synthesizing and non-synthesizing cyanobacteria: A bioinformatics study. Genomics. 2010;**95**:120-128. DOI: 10.1016/j.ygeno.2009.10.002

[90] Eisen JA, Wu M. Phylogenetic analysis and gene functional predictions: Phylogenomics in action. Theoretical Population Biology. 2002;**61**:481-487. DOI: 10.1006/tpbi.2002.1594

[91] Ward AC, Allenby NEE. Genome mining for the search and discovery of bioactive compounds: The Streptomyces paradigm. FEMS Microbiology Letters. 2018;**365**:1-20. DOI: 10.1093/femsle/ fny240

[92] Labreuche Y, Krin E, Ansquer D, Goudene D, Mangenot S, Calteau A, et al. Comparative genomics of pathogenic lineages of Vibrio nigripulchritudo identifies virulence-associated traits. The ISME Journal. 2013;**93**:1985-1996. DOI: 10.1038/ismej.2013.90

[93] Cuadrat RRC, Ionescu D, Dávila AMR, Marco DE. Recovering Genomics clusters of secondary metabolites from lakes using genome-resolved metagenomics. Frontiers in Microbiology. 2018;**9**:1-13. DOI: 10.3389/fmicb.2018.00251

[94] Kurita KL, Glassey E, Linington RG. Integration of high-content screening and untargeted metabolomics for comprehensive functional annotation of natural product libraries. PNAS. 2015; **112** p. 1 to 6. DOI: 10.1073/ pnas.1507743112

[95] Bhushan A, Egli PJ, Peters EE, Freeman MF, Piel J. Genome mining- and synthetic biology-enabled production of hypermodified peptides. Nature Chemistry. 2019;**11**:931-939. DOI: 10.1038/s41557-019-0323-9

[96] Machado H, Tuttle RN, Jensen PR. Omics-based natural product discovery and the lexicon of genome mining. Current Opinion in Microbiology. 2017; **39**:136-142. DOI: 10.1016/j. mib.2017.10.025

[97] Moreno-Pachon NM, Leeggangers HACF, Nijveen H, Severing E, Hilhorst H, Immink RGH. Elucidating and mining the Tulipa and Lilium transcriptomes. Plant Molecular Biology. 2016;**92**:249-261. DOI: 10.1007/ S11103-016-0508-1

[98] Nützmann H-W, Huang A, Osbourn A. Plant metabolic clusters—From genetics to genomics. The New Phytologist. 2016;**211**:771-789. DOI: 10.1111/NPH.13981

[99] Jiang C, Zhang H, Ren J, Dong J, Zhao X, Wang X, et al. Comparative transcriptome-based mining and expression profiling of transcription factors related to cold tolerance in peanut. International Journal of Molecular Sciences. 2020;**21**:1921. DOI: 10.3390/IJMS21061921

[100] Yadav R, Verma OP, Padaria JC. Transcript profiling and gene expression analysis under drought stress in *Ziziphus nummularia* (Burm.f.) Wright & Arn. Molecular Biology Reports. 2018;**45**: 163-174. DOI: 10.1007/S11033-018-4149-0

[101] Nath VS, Koyyappurath S, Alex TE, Geetha KA, Augustine L, Nasser A, et al. Transcriptome-based mining and expression profiling of Pythium responsive transcription factors in Zingiber sp. Functional &

Integrative Genomics. 2018;**19**:249-264. DOI: 10.1007/S10142-018-0644-6

[102] Sharma G, Aminedi R, Saxena D, Gupta A, Banerjee P, Jain D, et al. Effector mining from the *Erysiphe pisi* haustorial transcriptome identifies novel candidates involved in pea powdery mildew pathogenesis. Molecular Plant Pathology. 2019;**20**:1506-1522. DOI: 10.1111/MPP.12862

[103] Jongeneel CV. Searching the expressed sequence tag (EST) databases: Panning for genes. Briefings in Bioinformatics. 2000;**1**:76-92. DOI: 10.1093/BIB/1.1.76

[104] Lamblin A-FJ, Crow JA, Johnson JE, Silverstein KAT, Kunau TM, Kilian A, et al. MtDB: A database for personalized data mining of the model legume Medicago truncatula transcriptome. Nucleic Acids Research. 2003;**31**: 196-201. DOI: 10.1093/NAR/GKG119

[105] Kelso J, Visagie J, Theiler G, Christoffels A, Bardien S, Smedley D, et al. eVOC: A controlled vocabulary for unifying gene expression data. Genome Research. 2003;**13**:1222-1230. DOI: 10.1101/GR.985203

[106] Yonekura-Sakakibara K, Fukushima A, Saito K. Transcriptome data modeling for targeted plant metabolic engineering. Current Opinion in Biotechnology. 2013;**24**:285-290. DOI: 10.1016/J.COPBIO.2012.10.018

[107] Rao X, Dixon RD. Co-expression networks for plant biology: Why and how. Acta Biochimica et Biophysica Sinica. 2019;**51**:981-988. DOI: 10.1093/ABBS/GMZ080

[108] Fang X, Li C-Y, Yang Y, Cui M-Y, Chen X-Y, Yang L. Identification of a novel (−)-5-epieremophilene synthase from *Salvia miltiorrhiza* via transcriptome mining. Frontiers in Plant Science. 2017;**0**:627. DOI: 10.3389/FPLS.2017.00627

[109] Karunanithi PS, Dhanota P, Addison JB, Tong S, Fiehn O, Zerbe P. Functional characterization of the cytochrome P450 monooxygenase CYP71AU87 indicates a role in marrubiin biosynthesis in the medicinal plant Marrubium vulgare. BMC Plant Biology. 2019;**19**:1-14. DOI: 10.1186/S12870-019-1702-5

[110] Keeling CI, Weisshaar S, Ralph SG, Jancsik S, Hamberger B, Dullat HK, et al. Transcriptome mining, functional characterization, and phylogeny of a large terpene synthase gene family in spruce (Picea spp.). BMC Plant Biology. 2011;**11**: 1-14. DOI: 10.1186/1471-2229-11-43

[111] Aminfar Z, Rabiei B, Tohidfar M, Mirjalili MH. Identification of key genes involved in the biosynthesis of triterpenic acids in the mint family. Scientific Reports. 2019;**9**:1-15. DOI: 10.1038/s41598-019-52090-z

[112] Hodgson H, La Peña RD, Stephenson MJ, Thimmappa R, Vincent JL, Sattely ES, et al. Identification of key enzymes responsible for protolimonoid biosynthesis in plants: Opening the door to azadirachtin production. Proceedings of the National Academy of Sciences. 2019;**116**:17096-17104. DOI: 10.1073/PNAS.1906083116

[113] Singh A, Massicotte M-A, Garand A, Tousignant L, Ouellette V, Bérubé G, et al. Cloning and characterization of norbelladine synthase catalyzing the first committed reaction in Amaryllidaceae alkaloid biosynthesis. BioMed Cental Plant Biol. 2018;**18**:1-12. DOI: 10.1186/S12870-018-1570-4

[114] Hagel JM, Morris JS, Lee E-J, Desgagné-Penix I, Bross CD, Chang L, et al. Transcriptome analysis of 20 taxonomically related benzylisoquinoline alkaloid-producing plants. BMC Plant Biology. 2015;**15**:1-16. DOI: 10.1186/S12870-015-0596-0

[115] Lau W, Sattely ES. Six enzymes from mayapple that complete the biosynthetic pathway to the etoposide aglycone. Science. 2015;**349**: 1224-1228. DOI: 10.1126/SCIENCE. AAC7202

[116] Rizzo P, Altschmied L, Stark P, Rutten T, Gündel A, Scharfenberg S, et al. Discovery of key regulators of dark gland development and hypericin biosynthesis in St. John's wort (*Hypericum perforatum*). Plant Biotechnology Journal. 2019;**17**: 2299-2312. DOI: 10.1111/PBI.13141

[117] Pires ÁS, Rigueiras PO, Dohms SM, Porto WF, Franco OL. Structure-guided identification of antimicrobial peptides in the spathe transcriptome of the non-model plant, arum lily (*Zantedeschia aethiopica*). Chemical Biology & Drug Design. 2019;**93**:1265-1275. DOI: 10.1111/CBDD.13498

[118] Guerriero G, Piasecki E, Berni R, Xu X, Legay S, Hausman J-F. Identification of callose synthases in stinging nettle and analysis of their expression in different tissues. International Journal of Molecular Sciences. 2020;**21**:3853. DOI: 10.3390/ IJMS21113853

[119] Prasad P, Sreedhar RV. Identification and functional characterization of *Buglossoides arvensis* microsomal fatty acid desaturation pathway genes involved in polyunsaturated fatty acid synthesis in seeds. Journal of Biotechnology. 2019; **308**:130-140. DOI: 10.1016/J. JBIOTEC.2019.12.006

[120] Hellinger R, Koehbach J, Soltis DE, Carpenter EJ, Wong GK-S, Gruber CW. Peptidomics of circular cysteine-rich plant peptides: Analysis of the diversity of cyclotides from viola tricolor by transcriptome and proteome mining. Journal of Proteome Research. 2015;**14**: 4851-4862. DOI: 10.1021/ACS. JPROTEOME.5B00681

[121] Ran X, Zhao F, Wang Y, Liu J, Zhuang Y, Ye L, et al. Plant regulomics: A data-driven interface for retrieving upstream regulators from plant multi-omics data. The Plant Journal. 2020;**101**: 237-248. DOI: 10.1111/TPJ.14526

[122] Werner C, Doenst T, Schwarzer M. Metabolic pathways and cycles. In: The Scientist's Guide to Cardiac Metabolism. Amsterdam, The Netherlands: Elsevier Inc.; 2016. pp. 39-55. DOI: 10.1016/ B978-0-12-802394-5/00004-2

[123] Czarnecki JM, Shepherd AJ. Metabolic pathway mining. Methods in Molecular Biology. 2017;**1526**:139-158. DOI: 10.1007/978-1-4939-6613-4_8

[124] Rivera R, Garrido N. Metabolomics. In: Oxidants, Antioxidants, and Impact of the Oxidative Status in Male Reproduction. Amsterdam, Netherlands: Elsevier. 2018. pp. 277-285. DOI: 10.1016/B978-0-12-812501-4.00025-0

[125] Carbonell P. Enzyme discovery and selection. Metabolic Pathway Design: A Practical Guide. Berlin, Germany: Springer. 2019. pp. 63–81. DOI: 10.1007/ 978-3-030-29865-4_5

[126] Giraudeau P. NMR-based metabolomics and fluxomics: Developments and future prospects. Analyst. 2020;**145**:2457-2472. DOI: 10.1039/D0AN00142B

[127] Salon C, Avice JC, Colombié S, Dieuaide-Noubhani M, Gallardo K, Jeudy C, et al. Fluxomics links cellular functional analyses to whole-plant phenotyping. Journal of Experimental Botany. 2017;**68**:2083-2098. DOI: 10.1093/JXB/ERX126

[128] Singh S, Tiwari BS. Biosynthesis of high-value amino acids by synthetic biology. In: Current Developments in Biotechnology and Bioengineering: Synthetic Biology, Cell Engineering and

Bioprocessing Technologies. Amsterdam, Netherlands: DOI: 10.1016/B978-0-444-64085-7.00011-3

[129] Tian H, Zada B, Singh BH, Wang C, Kim SW. Synthetic biology approaches for the production of isoprenoids in *Escherichia coli*. In: Current Developments in Biotechnology and Bioengineering: Synthetic Biology, Cell Engineering and Bioprocessing Technologies. Amsterdam, Netherlands: Elsevier; 2018. Elsevier (Amsterdam, Netherlands). DOI: 10.1016/B978-0-444-64085-7.00013-7

[130] Saini DK, Pabbi S, Prakash A, Shukla P. Synthetic biology applied to microalgae-based processes and products. In: Handbook of Microalgae-Based Processes and Products. Amsterdam, The Netherlands: Elsevier Inc.; 2020. DOI: 10.1016/b978-0-12-818536-0.00004-x

[131] Feiner A, Pitra N, Matthews P, Pillen K, Wessjohann LA, Riewe D. Downy mildew resistance is genetically mediated by prophylactic production of phenylpropanoids in hop. Plant, Cell & Environment. 2021;**44**:323-338. DOI: 10.1111/PCE.13906

[132] Ranganathan S, Zhao Y, Simon R. Encyclopedia of Systems Biology. Berlin, Germany: Springer; 2013. DOI: 10.1007/978-1-4419-9863-7

[133] Vick JE, Johnson ET, Choudhary S, Bloch SE, Lopez-Gallego F, Srivastava P, et al. Optimized compatible set of BioBrick™ vectors for metabolic pathway engineering. Applied Microbiology and Biotechnology. 2011; **92**:1275-1286. DOI: 10.1007/s00253-011-3633-4

[134] Camp D, Garavelas A, Campitelli M. Analysis of Physicochemical Properties for Drugs of Natural Origin. Journal of Natural Products. 2015;**78**:1370-1382. DOI: 10.1021/acs.jnatprod.5b00255

[135] Haustedt LO, Siems K. The role of natural products in drug discovery: Examples of marketed drugs. In: Small Molecule Medicinal Chemistry: Strategies and Technologies. Hoboken, US: John Wiley & Sons, Inc.; 2016. pp. 381-430.

[136] Newman DJ, Cragg GM. Natural products as sources of new drugs over the nearly four decades from 01/1981 to 09/2019. Journal of Natural Products. 2020;**83**:770-803. DOI: 10.1021/ACS.JNATPROD.9B01285

[137] Wessjohann LA. Reverse metabolomics—Metabolomics in drug discovery: Connecting metabolomic profiles with phylogenetic, medicinal and flavoring properties. Metabolomics. 2014;**s1**. DOI: 10.4172/2153-0769.S1.024

[138] Zhou JZ. Chemoinformatics and library design. In: Chemical Library Design: Methods and Protocols. Berlin, Germany: Springer; 2011. pp. 27-52. Springer (Berlin, Germany). DOI: 10.1007/978-1-60761-931-4

[139] Santana K, do Nascimento LD, Lima e Lima A, Damasceno V, Nahum C, Braga RC, et al. Applications of virtual screening in bioprospecting: Facts, shifts, and perspectives to explore the chemo-structural diversity of natural products. Frontiers in Chemistry. 2021;**9**. Article: 662688. DOI: 10.3389/fchem.2021.662688

[140] Reymond J-L, Awale M. Exploring chemical space for drug discovery using the chemical universe database. ACS Chemical Neuroscience. 2012;**3**:649-657. DOI: 10.1021/cn3000422

[141] Liu R, Li X, Lam KS. Combinatorial chemistry in drug discovery. Current Opinion in Chemical Biology. 2017;**38**: 117-126. DOI: 10.1016/j.cbpa.2017.03.017

[142] Wessjohann LA. Synthesis of natural-product-based compound

libraries. Current Opinion in Chemical Biology. 2000;**4**:303-309. DOI: 10.1016/S1367-5931(00)00093-4

[143] Kodadek T. The rise, fall and reinvention of combinatorial chemistry. Chemical Communications. 2011;**47**: 9757-9763. DOI: 10.1039/c1cc12102b

[144] Schulz MN, Hubbard RE. Recent progress in fragment-based lead discovery. Current Opinion in Pharmacology. 2009;**9**:615-621. DOI: 10.1016/j.coph.2009.04.009

[145] Erlanson DA, Fesik SW, Hubbard RE, Jahnke W, Jhoti H. Twenty years on: The impact of fragments on drug discovery. Nature Reviews. Drug Discovery. 2016;**15**: 605-619. DOI: 10.1038/nrd.2016.109

[146] Chen H, Zhou X, Wang A, Zheng Y, Gao Y, Zhou J. Evolutions in fragment-based drug design: The deconstruction-reconstruction approach. Drug Discovery Today. 2015; **20**:105-113. DOI: 10.1016/j.drudis.2014.09.015

[147] Ahmed J, Worth CL, Thaben P, Matzig C, Blasse C, Dunkel M, et al. FragmentStore—A comprehensive database of fragments linking metabolites, toxic molecules and drugs. Nucleic Acids Research. 2011;**39**: 1049-1054. DOI: 10.1093/nar/gkq969

[148] Scott DE, Coyne AG, Hudson SA, Abell C. Fragment based approaches in drug discovery and chemical biology. Biochemistry. 2012;**51**:4990-5003. DOI: 10.1021/bi3005126

[149] Keserü GM, Makara GM. The influence of lead discovery strategies on the properties of drug candidates. Nature Reviews. Drug Discovery. 2009; **8**:203-212. DOI: 10.1038/nrd2796

[150] Vásquez AF, González BA. Pushing the ligand efficiency metrics: Relative group contribution (RGC) model as a helpful strategy to promote a fragment "Rescue" effect. Frontiers in Chemistry. 2019;**7**:1-7. DOI: 10.3389/fchem.2019.00564

Section 2

Engineering Genomes through CRISPR-Based Technologies and Mathematical Modelling

Chapter 3

Applications of CRISPR/Cas Technology to Research the Synthetic Genomics of Yeast

Huafeng Lin, Haizhen Wang, Aimin Deng, Minjing Rong, Lei Ye, Lei Shi, Tuanmei Wang and Xiangwen Peng

Abstract

The whole genome projects open the prelude to the diversity and complexity of biological genome by generating immense data. For the sake of exploring the riddle of the genome, scientists around the world have dedicated themselves in annotating for these massive data. However, searching for the exact and valuable information is like looking for a needle in a haystack. Advances in gene editing technology have allowed researchers to precisely manipulate the targeted functional genes in the genome by the state-of-the-art gene-editing tools, so as to facilitate the studies involving the fields of biology, agriculture, food industry, medicine, environment and healthcare in a more convenient way. As a sort of pioneer editing devices, the CRISPR/Cas systems having various versatile homologs and variants, now are rapidly giving impetus to the development of synthetic genomics and synthetic biology. Firstly, in the chapter, we will present the classification, structural and functional diversity of CRISPR/Cas systems. Then we will emphasize the applications in synthetic genome of yeast (*Saccharomyces cerevisiae*) using CRISPR/Cas technology based on year order. Finally, the summary and prospection of synthetic genomics as well as synthetic biotechnology based on CRISPR/Cas systems and their further utilizations in yeast are narrated.

Keywords: applications, CRISPR/Cas, gene editing, *S. cerevisiae*, synthetic genomics, yeast

1. Introduction

Synthetic biology is a fundamentally interdiscipline. It has become an important methodology in biotechnology owing to its novel functions and regulation mechanisms. The scientific concept of synthetic biology can be traced back to the book "The Mechanism of Life" written by a French physical chemist Stéphane Leduc in 1911 [1]. It currently refers to the practical application discipline that integrating modern science and engineering technology to promote and accelerate the design, alteration and creation of bio-genetic materials in living organisms [2]. Briefly, synthetic biology can roughly be considered as the reverse process of analytical biology. Sustained advances in synthetic biology will depend on coordinated and paralleled developments within many different discipline areas, and cooperation of scientists from most countries. In China, scientists have made landmark contributions in this field with the success of artificial synthesis of bovine insulin and yeast (*S. cerevisiae*) alanine transfer RNA. In USA, artemisinic acid, a precursor of antimalarial drug artemisinin, was synthesized in *S. cerevisiae* with a yield of 25 g/L, and has been industrialized in 2013 [3]. Nowadays, synthetic biology is rapidly penetrating into various fields including bio-science, gene engineering, agriculture, food industry and medicine, in which synthetic genomics plays a profound role in providing theoretical basis and technological support. Synthetic genomics is viewed as an important area of synthetic biology, which being engineered under a general genome scope, mainly refers to design and assembly of nucleotide fragments to generate functional living genomes [4], including recreated and recoded genomes as well as minimal genomes. Genetic manipulation is known as one of the central strategies to investigate the molecular basis of living things as well as their evolution and diversity, which advancing the understanding of biological systems at a micro level. Compared to conventional approaches in genetic manipulation, synthetic genomics has the characteristics of introducing large numbers and diversity of genetic modifications [5]. Additionally, synthetic genomics can theoretically create a synthetic genome to practically and feasibly build a simpler and more amenable genome-scale platform for biological system construction [6].

In a certain sense, the first synthetic gene synthesized in 1970s [7] marked the beginning of synthetic genomics. Then, viral chromosomes were the first to be synthesized in the early 2000's because of their comparatively small size [8]. Nowadays, as a first designer synthetic eukaryotic genome, Sc2.0 (*S. cerevisiae* 2.0) project has achieved significant progresses including real-world applications for industrial microbiology, and may create a big economic value in the future. Driven by rapid advances in gene assembly, genome editing and mathematical modeling techniques, synthetic genomics is developing quickly. As a young discipline, synthetic genomics has helped to promote our new understanding of genome structure and function. Recently, an important direction of synthetic genomics is to transform the natural biological systems through gene editing techniques. With emerging novel classes of programmable genetic tools, in particular, the establishment and optimization of CRISPR and associated technology platforms, synthetic biology and its vital field——synthetic genomics is entering a new era of more possibilities. Actually, early in 2014, the European Commission's synthetic biology summit has typically categorized synthetic biology tools as design, construction and diagnostic tools whereas synthetic biology methodology serves the study of DNA synthesis and synthetic genomics, engineering biology, xenobiology as well as protocell biology [9]. CRISPR/Cas nucleases have been extensively applied to manipulate the genomes of cultured and primary cells, animals and plants, vastly accelerating the pace of basic research and enabling breakthroughs in the field of synthetic biology and synthetic genomics [10]. It can be expect that, CRISPR toolkits are of particular importance to the future of synthetic genomics due to its great potential to open new pathways for manipulation and expression of genetic information, which will certainly transform synthetic genomics and synthetic biology greatly.

Here in this chapter, we review the developments of CRISPR/Cas technology, the main types of CRISPR/Cas system, as well as the applicational research of synthetic genomics in yeast using CRISPR/Cas toolboxes. Finally, we also provide perspectives on future directions and applications of CRISPR/Cas-based methodology in the research of synthetic yeast genome.

2. CRISPR/Cas: a powerful and versatile toolkits for synthetic genomics

The survival battle between microbes and bacteriophage is the driving force behind the evolution and diversification of microbial adaptive immune system.

As part of the immune response in bacteria, CRISPR/Cas systems are responsible for tackling the invading phages or plasmids. These systems are of particular importance to the future of synthetic genomics owing to their great potential to open new doors for manipulation and expression of genetic sequences. Since the discovery of CRISPR in the *Escherichia coli* genome in 1987 [11], CRISPR/Cas systems now have been classified into two main classes (Class 1 and Class 2) and six types (type I, II, III, IV, V and VI) covering over 30 different subtypes [12]. The diversities of architecture and classification render CRISPR/Cas systems with broad functional versatilities. Of note, the diversity of protospacer adjacent motif (PAM) is also an important feature for CRISPR/Cas systems. Briefly, PAM refers to a short sequence resides in the exogenous nucleic acid elements (commonly at the 3′ end of the target DNA) but not the CRISPR array and its guide RNAs that support to discriminate self versus non-self of microbes' nucleic acid ingredients. The nucleotide sequences that can be edited by CRISPR/Cas systems have been limited by the PAM and gRNA sequences. Commonly, *Streptococcus pyogenes* Cas9 (SpCas9) recognizes 5'-NGG-3′, and *Staphylococcus aureus* Cas9 (SaCas9) recognizes 5'-NNGRRT-3′ as their respective functional PAM [13]. However, type V CRISPR/Cas system uses a T-rich PAM sequence such as "TTT", "TTA", "TTN" or "TTC" for target recognition. For example, Cas12a (Cpf1) uses the "TTN" PAM sequence to target dsDNA [14]. Cas proteins and their variants require different PAM sequences for efficiently and precisely manipulate and cleavage biological genomes, while type VI system relies on a PAM analogue termed protospacer flanking site (PFS) for RNA targeting [15]. Moreover, taking Cas9 protein as an example, a previous study has reported the important function of PAM recognition in the field of inducing target DNA unwinding, which underscored by the sequential variability of PAM recognition presented in engineered Cas variants [16].

In addition to the basic properties for antiviral roles, CRISPR/Cas systems have numerous potential applications for gene editing, transcriptional activation/inhibition, epigenetic modification, chromatin imaging, single base substitution, point-of-care diagnostics [17] and synthetic biology, among others [18]. However, taking account of off-target concerns, much more interrogations are needed to make these systems valuable and reliable toolboxes for utility research area.

As CRISPR systems continue to be discovered, a variety of programmable nucleases have joined the ranks of genome editing. Currently, three types of candidate nuclease systems including Cas9 and its homologs, DNA-targeting Cas12, as well as RNA-targeting Cas13 have together become the focus of gene-editing technology. Cas9 nuclease induces double strand breaks (DSB), and Cas9 nickase triggers single strand breaks (SSB); whereas dCas9 (dead Cas9) derived from Cas9 nuclease via mutating the HNH nuclease domain and RuvC-like domain, can still bind to the target region for transcriptional interventions, but without introducing a DSB. Based on similar mechanisms, dCas12a and dCas13 are capable to mediate CRISPR interference (CRISPRi) and CRISPR activation (CRISPRa) at the transcriptional levels. CRISPRi/a is a tremendously useful tool for transcriptional regulations which achieving by covalent modification of histones of Cas proteins as it allows to balance and optimize gene expression without genome editing. Besides, Cas9 and Cas12a (previously named as Cpf1) are RNA-guided endonucleases that can induce genome editing by triggering DSB repair at a specific site [19]. Cas9 and Cas13a can be used for targeted RNA interference. Cas13a fused to base editors can be used to modify nucleotides in RNA. Futhermore, the dCas nucleases have termed "discovery-based synthetic biology", constitute one part of toolbox to study synthetic biology and engineer biology, especially have functions in the field of characterizing the function of noncoding genes and regulatory elements of the genome, and strategies to design synthetic gene circuits [20]. Theoretically, CRISPR toolkits can be applied to integrate synthetic sequences into the genome, to form genomic libraries, and to target on multiple functional loci conveniently for gene modifications with high efficiency in strains or cell lines. The increased knowledges of CRISPR classifications and their action mechanisms open up new scopes of applications in synthetic genome of an organism via these systems.

3. Synthetic genomics and the use of CRISPR technology in synthetic genomics of *S. cerevisiae*

Technical advances in chemical synthesis of poliovirus cDNA in the absence of a natural template have gained global attentions. The successful synthesis of the 7,740 bp poliovirus cDNA artificially not only indicated the feasibility of producing infectious virus using chemically synthesized oligonucleotides as initial materials but also showed the great potential of modifying and creating more complex genomes under laboratory conditions [21]. Then in 2003, the artificial synthesis of ΦX174 bacteriophage genome (5,386 bp) using synthetic oligonucleotides has paved the way for synthesizing bacterial genomes [22]. Since then, several other viral genomes and transposons have also been synthesized in their entirety. Bacterial genomes comprising millions of base pairs are very complex, which often regarded as advanced factories to synthetise biologically active chemical substances. Synthetic biologist Craig Venter and his research team have successively synthesized the 582,970 bp *Mycobacterium genitalium* genome and *Mycoplasma mycoides* genome JCVI-1.0, and the latter represented a milestone in the history of synthetic genomics [23, 24].

From historical perspective of human domestication, potentially the next best thing to fire is yeast [25]. In practical terms, yeast represents one of the simplest eukaryotic microorganisms, and as an attractive model organism has been widely used in the food industry [26]. The Synthetic Yeast Genome (Sc2.0) Project is ongoing and aims to explore yeast chromosome structure, minimal eukaryotic genome length and gene content, as well as to rewrite all 16 yeast chromosomes. As we known, DNA synthesis is an essential tool for synthetic genomics. Currently, with the aid of CRISPR/Cas technology, for instance, programmable Cas9 proteins can execute sequence-specific DSB depending on the target sites of the gRNA, and also genes in the genome of an organism can be programmed and are rewritable [27]. Generally, CRISPR/Cas systems have vastly simplified genome editing in yeasts via performing gene over-expression, knockin, knockout, mutations and deletions, and enabled easy-operation genetically engineering of products of fuel molecules, chemical components, food ingredients, and active pharmaceutical ingredients.

Early in 2013, DiCarlo and colleagues firstly piloted CRISPR/Cas9 system to engineer for site-specific mutagenesis and allelic replacement in the genome of *S. cerevisiae* using dsOligo (double-stranded 90-bp oligonucleotide) as a template with efficiency rates close to 100% [28]. A few months later, Farzadfard's research team reported a CRISPR/Cas-based eukaryotic transcriptional regulation system implemented in *S. cerevisiae*, which will open up new paths for drawing natural genetic circuits and their regulations regarding on cellular phenotypic mechanisms [29]. In the years following these initial works, several related research reports have been emerging.

In 2014, Ryan and others improved the utilization of fiber disaccharide in diploid yeast by multiplex CRISPR system, which made the cellobiose fermentation rates increased by more than 10 times [30]. Similarly as DiCarlo's experimental methodology, Zhang et al. engineered the industrial polyploid strain ATCC4124, where URA3, TRP1, LEU2 and HIS3 were knocked out one-by-one with efficiencies varying from 15–60% to create an auxotrophic strain. And this method is likely to be very valuable for yeast genome engineering due to having no need of selectable markers that labeled in the integrated DNA [31].

In 2015, Bao et al. and Mans et al. respectively, used a HI-CRISPR (homology-integrated CRISPR) or CRISPR/Cas9 strategy to successfully generate multiple gene modifications in yeast *S. cerevisiae* simultaneously [32, 33]. In the studies related to metabolic pathway engineering, Ronda et al. applied a new system called CrEdit (CRISPR/Cas9 mediated genome Editing) to enable simultaneous and highly efficient integration of three pathway genes (BTS1, crtYB and crtI) involved in the production of β-carotene at three different integration sites (X-3, XI-2, and XII-5 gene locus) in the genome of *S. cerevisiae* [34]. Jakočiūnas and collaborators applied CRISPR/Cas9 for multiplex gene knock-out to search for strains with improved production of mevalonate (a key intermediate for isoprenoid and sterol production) in yeast [35], followed by the utilization of CRISPR/Cas9 system to integrate crtYB, CrtI and crtE genes in three gene sites (ADE2, HIS3 and URA3) of *S. cerevisiae* for successfully constructing carotenoid biosynthesis pathway [36].

In 2016, more studies in relation to the use of CRISPR/Cas9 system for engineering in the genome of yeast have been reported [37–44]. Most notably, Tsarmpopoulos et al. reported the CRISPR/Cas9 adaptation for the engineering of bacterial genomes cloned in yeast. The result showed that applying 90 nt paired oligonucleotides as templates to promote recombination which achieved a seamless deletion of the mycoplasma glpO (glycerol-3-phosphate oxidase-encoding) gene without selection in one step. This work paves the way to high-throughput manipulation of natural or synthetic genomes in yeast *S. cerevisiae* [37].

In the year of 2017, Vanegas and her partners used a combination tool of Cas9 genome editing and dCas9 transcriptional regulation to engineer *S. cerevisiae* for production of flavonoid precursor naringenin and simultaneously restrainting formation of by-product phloretic acid [45]. Reider Apel et al. constructed a clone-free toolkit based on CRISPR/Cas9, which solved the problems of chromosome integration locus and promoter selection, protein localization and solubility in yeast metabolic engineering, and optimized the expression of taxadiene synthase by using the tool, which increased the yield of taxadiene by 25 times [46]. Contrary to the result of protein overexpression, Vigentini et al. employed the CRISPR/Cas9 system to successfully reduce urea production in *S. cerevisiae* wine yeasts via eliminating the CAN1 arginine permease pathway [47]. Interestingly, Mans et al. used CRISPR/Cas9 technology to explore the elusive mechanism for lactate export in *S. cerevisiae*. The 25-deletion strain in this experiment has taken the first step in building a yeast's 'minimal transportome' platform, which can be applicable to functional explanation of heterologous transport proteins and the assessment of metabolic engineering strategies [48]. The summary of CRISPR-based studies in yeast in the year of 2017 have showed in **Table 1** [45–59].

In the beginning of the year 2018, Verwaal et al. employed three gene-editing systems, Cpf1 orthologues (*Acidaminococcus spp*. BV3L6 (AsCpf1), *Lachnospiraceae bacterium* ND2006 (LbCpf1) and *Francisella novicida* U112 (FnCpf1)) for genome modification of *S. cerevisiae*. The result of this work demonstrated that Cpf1 can broaden application sphere of the genome-editing toolbox available for research of *S. cerevisiae* [69]. Li et al. firstly used the CRISPR/Cpf1 to delete large DNA fragment (the deletion of DNA fragment of ~38 kb between the two genes of TRM10 and REX4) in *S. cerevisiae*, which demonstrating that the CRISPR/Cpf1 system can be used for genome simplification of *S. cerevisiae*, and to facilitate the laboratory evolution of the genome of *S. cerevisiae* [70]. Later in the year, Dank et al. used CRISPR/Cas9 technology to construct *S. cerevisiae* mutants with lacking esterase IAH1 and/or TIP1. Very interestingly, not affecting by the double gene knockout of yeast mutant ΔIAH1ΔTIP1, a complex regulatory mechanism to compensate

Gene Action Modes	CRISPR System	Expression Products	Gene Sites (or gene number)	Authors and Year	References
Downregulation	Cas9/dCas9 based system	Naringenin	*TSC*13	Vanegas *et al.* (2017)	[45]
Gene integration	CRISPR/Cas9	Taxadiene	23 genomic loci	Reider Apel *et al.* (2017)	[46]
Genetic modification	CRISPR/Cas9	Urea	*CAN*1	Vigentini *et al.* (2017)	[47]
Gene deletion	CRISPR/Cas9	Lactate	25 genomic loci	Mans *et al.* (2017)	[48]
Gene regulation	dCas9-VPR	Naringenin	*NDE*2, *CYC*1, *GPD*1, *TDH*1	Vanegas *et al.* (2017)	[49]
Gen regulation and integration	CRISPR/Cas9	Cellulase; isobutanol; glycerol	δ-regions	Si *et al.* (2017)	[50]
Graded expression	CRISPR/dCas9	Pathway enzymes	Multi-genes (e.g., *ZWF*1, *TAL*1, *TKL*1)	Deaner *et al.* (2017)	[51]
Gene modulation	dCas9-VPR	-	Up to 4 native genes	Deaner *et al.* (2017)	[52]
Transcriptional reprogramming	dCas9 systems	Isoprenoid; TAG	-	Jensen *et al.* (2017)	[53]
Logic circuits	dCas9-Mxi1	-	-	Gander *et al.* (2017)	[54]
Gene regulation	CRISPR-AID	Beta-carotene	*HMG*1; *ERG*9; *ROX*1	Lian *et al.* (2017)	[55]
Gene editing	CRISPR/Cas9	Glutathione	*ADE*2, *URA*3, *LEU*2, *TRP*1, *HIS*3	Zhou *et al.* (2017)	[56]
Construction of mutants	"CRISPR Nickase system"	-	*CAN*1, *CDC*25	Satomura *et al.* (2017)	[57]
Strain Generation; Gene drive	CRISPR/Cas9	Strain mutants	-	Roggenkamp *et al.* (2017)	[58]
Genome editing; point mutation	FnCpf1	-	*ADE*2, *HIS*4, *PDR*12, *CAN*1	Swiat *et al.* (2017)	[59]
Multiplexed engineering	GTR-CRISPR	-	8 genes (e.g., *CAN*1, *ADE*2, *LYP*1, etc.)	Ferreira *et al.* (2018)	[60]
Gene activation, interference, and deletion	dLbCpf1, dSpCas9, SaCas9	Recombinant *Trichoderma reesei endoglucanase* II	Unspecified target genes	Schultz *et al.* (2018)	[61]
Chromosome fusion	CRISPR/Cas9	-	Multiple-chromosome	Shao *et al.* (2018)	[62]
Genomic integration	CRISPR/Cas9	Natural genetic variants	*SEC*14 gene	Roy *et al.* (2018)	[63]
Gene disruption	CRISPR/Cas9	Bioethanol	*ADH*2	Xue *et al.* (2018)	[64]

Gene Action Modes	CRISPR System	Expression Products	Gene Sites (or gene number)	Authors and Year	References
Cocktail integration	CRISPR/Cas9 combination	Target products: beta-carotene	*ADE2*, *URA3*, *CAN1*	Hou *et al.* (2018)	[65]
Genetic manipulations	CRISPR/Cas9	Mating-types, diploids and polyploids	*MAT* locus	Xie *et al.* (2018)	[66]
Single-nucleotide genome-editing	CRISPR/Cas9 combination	Genetic variants	*CAN1*, *ADE2*, *LYP1*, etc.	Bao *et al.* (2018)	[67]
Genomic integration	FnCpf1	Beta-carotene	*Gal1-7* locus, *Gal80* locus, *HO* locus, etc.	Li *et al.* (2018)	[68]

Table 1.
The summary table of CRISPR-based studies in yeast in the year of 2017 and 2018.

multiple genomic defects in aroma metabolism is generated and activated to show an aroma composition comparable to wild type levels [71]. Using CRISPR technology, the related studies in yeasts for creating genome mutations and integrations have also been respectively carried out by Guo et al. [72] and Jakociunas et al. [73]. And much more related works have listed in **Table 1** [60–68, 74].

As CRISPR systems continue to be extensively used, dozens of articles (more than 60 scientific papers in PubMed) in relation to CRISPR-based yeast have been published in 2019. Based on PubMed database, 2019 is the year of the most articles published in recent years in the field of CRISPR-edited yeasts. Among them, Zhang et al. deleted 6 genes in the yeast genomes in 3 days through developing a multiplexed gene-editing platform termed GTR-CRISPR (a gRNA-tRNA array for CRISPR/Cas9) with 60% efficiency using reported gRNAs and 23% using unimproved gRNAs. They further concluded that GTR-CRISPR may be suggested to the most valuable complement for the toolkit of synthetic biology and auto-operation [75]. Notably, Laughery et al. found that the cause of dCas9 targeting boosted mutagenesis in yeast is likely affected mutationally by dCas9-mediated R-loop formation. These findings not only showed important implications for the applications of additional mutagenesis in dCas9 (and Cas9), but also offered a novel method for interrogating the mechanism of targeted R-loop formation induces eukaryotic genome variability and/or mutagenesis [76]. As excellent as above two work, another two investigations respectively described the CRISPR/Cas9 based functional chromosome fusions [77] and CRISPR/dCas9 based AND gate in yeast [78]. Additionally, as typical research examples, this review only list 5 valuable articles published in 2019 (**Table 2**) [79–83].

Good works are also seen in 2020. Wu and his team found that specific cleavage via CRISPR/Cas9 near the centromere of a *S. cerevisiae* chromosome can lead to elimination of the whole chromosome and initiate chromosome drive [89]. van Wyk et al. utilized CRISPR/Cas9 technology to generate a self-cloned wine yeast strain that over-expresses two genes of oenological relevance, GPD1 (glycerol-3-phosphate dehydrogenase 1) and ATF1 (alcohol acetyltransferase 1), which directly implicated in glycerol and acetate ester production respectively. The result provided an alternative strategy to obtain increased glycerol and lower acetic acid levels, without disrupting the aldehyde dehydrogenase activity [90]. In addition to the above two, here enumerate 5 representative literatures on CRISPR-based yeasts in this year used for references (**Table 2**) [84–88].

References	Authors/year	Gene regulation	Cas modes	Target loci	Descriptions
[79]	Ai et al./2019	Deletion	Cas9	gal80	Engineered yeasts that can produce artemisinic acid without galactose induction.
[80]	Chen et al./2019	down-regulation or knock-out	Cas9	erg9; rox1; ypl062w; yjl064w	Achieved high production of valencene through CRISPR/Cas9-mediated metabolic engineering.
[81]	Laughery et al./2019	Genome editing	Cas9	RNR1 gene	Describe a simple protocol for constructing Cas9-expressing plasmids and protocols for genome editing in yeast.
[82]	Ciurkot et al./2019	Genome editing	Cas12a	INT1, INT2 and INT3 genomic sites	Demonstrated Cas12a-mediated multiplex genome editing in yeasts, and created yeast pixel art with an acoustic liquid handler using yeast strains that producing differently colored carotenoid.
[83]	Yan & Finnigan/2019	Gene drive	Cas9	HIS3 locus	Developed a gene drive in yeasts that allowing for the examination of alternative drive designs and control mechanisms.
[84]	Cámara et al./2020	Gene expression	CRISPR/Cas	PDR12 locus	Constructed a toolkit for CRISPRa/i for a polyploid industrial yeast strain.
[85]	Levi & Arava/2020	Gene modification	Cas9	HTS1; OM14; FRS1	Provided a facile alternative to manipulate the yeast genome.
[86]	Li et al./2020	Genome engineering	Cas9	UPC2 gene; Gal1-7, Gal80 sites	Cas-3P allowed single-, double- and triple-loci gene targeting in yeasts.
[87]	Mitsui et al./2020	Genome evolution	GMES /CRISPR	13 genes: HXT7, HXK2, PGI1, etc.	Aimed to construct a lactic acid-tolerant yeast to reduce the neutralization cost in LA production.
[88]	Yang et al./2020	Gene targeted mutation	Cas9	ERG20	Facilitate construction of genomic mutations of essential genes for functional genomic analysis and metabolic flux regulation in yeasts.

Note: CRISPRa/i, CRISPR activation and interference; Cas-3P, Cas9 and three marked plasmid backbones; GMES/CRISPR, global metabolic engineering strategy/CRISPR.

Table 2.
CRISPR-based applications in yeast in the year of 2019.

Stepping into 2021, innovations remain advancing. Gong et al. reported a gRNA-tRNA array and SpCas9-NG (GTR 2.0) for the applications of highly efficient genome damage and base editing. In this study, they achieved gene cleavages with almost 100 efficiencies in the cells of *S. cerevisiae*. During the process, gene editing includes all 16 possible NGN PAMs and all 12 possible single-nucleotide mutations (N to N). Further, they employed GTR 2.0 system for multiplexed single-nucleotide mutations to simultaneously produce 4 single-nucleotide conversions in single gene with 66.67% mutation efficiency, and to create simultaneously 2 single-nucleotide mutations in two different genes with 100% mutation efficiency [91].

McGlincy et al. showed a comprehensive yeast CRISPRi library, based on empirical design rules, containing 10 distinct guides for most genes, which providing a strategy for genome-wide CRISPR interference screening in budding yeast [92]. Furthermore, a short communication introduced a GDi-CRISPR system (gene drive delta site integration system by the CRISPR system) for multi-copy integration in *S. cerevisiae*, which holds great promising for advancing the development of *S. cerevisiae* multi-copy integration tools [93].

4. Future challenges and prospections in the applications

As an emerging field, synthetic biology has high potential applications in drug-discovery, development of medical therapeutics, diagnostic tools and improvement of bioproducts. And its emerging applications include vaccine development, cancer treatment, prevention and treatment of infection, microbiome engineering, cell therapy and regenerative medicine, biofuels as well as genome engineering [94]. New technologies, such as CRISPR/Cas-mediated genome editing, will enable synthetic biologists to take a more holistic engineering approach, modifying synthetic circuits and the host genome with relative ease [95]. Nowadays, the CRISPR/Cas system is only 8 years old. With great progress in gene editing technology, CRISPR/Cas systems surely will greatly boost the development of gene therapy, basic biological research, and synthetic biology, let alone in the research field of *S. cerevisiae*. However, the applications of CRISPR/Cas systems have still encountered several major challenges including off-target effects, delivery modalities, Cas9 cleavage activities and immune responses.

Off-target effects of CRISPR/Cas systems usually result from mismatches between the guide RNAs and their target gene sequences [96], and may result in targeting to unexpected sequences of nucleic acids. Many efforts have been done to lower unwanted off-target effects of CRISPR/Cas. Previous studies reported that the systems could effectively maximize on-target activity and minimize off-target effects for genome engineering either by modifying guide RNA or Cas9 nuclease [97, 98].

5. Conclusion

Thanks to the eximious predominance of yeast genetics, the organism *S. cerevisiae* has enjoyed gene regulation technology for decades. In spite of some limitations needed to be conquered, the advent of the CRISPR techniques have undoubtedly created a new era for genomic engineering in yeasts. CRISPR/Cas system as one of the most cutting-edge biotechnology will continue to not only improve our insight into the complexity of cells but also help us make better utilization of living systems. Taking together, the use of CRISPR/Cas systems for various synthetic biology applications, specially in the synthetic genome of yeast *S. cerevisiae*, has greatly accelerate food industry, biomedical study and agricultural research.

Acknowledgements

This work was collectively supported by grants from National key research and development plan (2016YFD0500600), Guangdong provincial science and technology plan project (2017B020207004), Innovative province construction project--special topic on fighting against novel coronavirus pneumonia epidemic (2020SK3044), Youth fund project of Hunan natural science foundation (2019JJ50681) and Changsha biological resources sample bank establishment project (20200365).

Conflict of interest

The authors declare no conflict of interest.

Author details

Huafeng Lin[1,2,3†], Haizhen Wang[4†], Aimin Deng[1], Minjing Rong[2], Lei Ye[2], Lei Shi[2], Tuanmei Wang[1*] and Xiangwen Peng[1*]

1 Changsha Hospital for Maternal and Child Health Care of Hunan Normal University, Jinan University, Changsha, China

2 Institute of Food Safety and Nutrition, Jinan University, Guangzhou, China

3 Department of Biotechnology, College of Life Science and Technology, Jinan University, Guangzhou, China

4 Department of Life Science, Lu Liang University, Luliang, China

*Address all correspondence to: 1396308863@qq.com and pxw1237@163.com

† These authors contributed equally to this work.

References

[1] Tirard S. (2009). Stéphane Leduc (1853-1939), de la médecine à la biologie synthétique [Stephane Leduc (1853-1939), from medicine to synthetic biology]. Histoire des sciences medicales, 43(1), 67-72.

[2] Heidari, R., Shaw, D.M., & Elger, B.S. (2017). CRISPR and the Rebirth of Synthetic Biology. Science and engineering ethics, 23(2), 351-363. doi: 10.1007/s11948-016-9768-z

[3] Paddon, C.J., Westfall, P.J., Pitera, D.J., Benjamin, K., Fisher, K., McPhee, D., et al. (2013). High-level semi-synthetic production of the potent antimalarial artemisinin. Nature, 496(7446), 528-532. doi: 10.1038/nature12051

[4] Kumar, J., Narnoliya, L. K., & Alok, A. (2019). A CRISPR Technology and Biomolecule Production by Synthetic Biology Approach (pp. 143-161). doi: 10.1016/B978-0-444-64085-7.00006-X

[5] Coradini, A., Hull, C.B., & Ehrenreich, I.M. (2020). Building genomes to understand biology. Nature communications, 11(1), 6177. doi: 10.1038/s41467-020-19753-2

[6] Chen, B., Lee, H. L., Heng, Y. C., Chua, N., Teo, W. S., Choi, W. J., Leong, S., Foo, J.L., & Chang, M.W. (2018). Synthetic biology toolkits and applications in *S. cerevisiae*. Biotechnology advances, 36(7), 1870-1881. doi: 10.1016/j.biotechadv.2018.07.005

[7] Schindler, D., Dai, J., & Cai, Y. (2018). Synthetic genomics: a new venture to dissect genome fundamentals and engineer new functions. Current opinion in chemical biology, 46, 56-62. doi: 10.1016/j.cbpa.2018.04.002

[8] Wimmer, E., Mueller, S., Tumpey, T.M., & Taubenberger, J.K. (2009). Synthetic viruses: a new opportunity to understand and prevent viral disease. Nature biotechnology, 27(12): 1163-1172. doi: 10.1038/nbt.1593

[9] European Commission. (2014). Opinion on synthetic biology I definition. http://ec.europa.eu/health/scientific_committees/emerging/docs/scenihr_o_044.pdf. Accessed 07 Dec 2015.

[10] Doudna, J.A. (2020). The promise and challenge of therapeutic genome editing. Nature, 578: 229-236. doi: 10.1038/s41586-020-1978-5

[11] Ishino, Y., Shinagawa, H., Makino, K., Amemura, M. and Nakata, A. (1987). Nucleotide sequence of the IAP gene, responsible for alkaline phosphatase isozyme conversion in *E. coli*, and identification of the gene product. Journal of bacteriology, 169(12): 5429-5433. doi: 10.1128/jb.169.12.5429-5433.1987

[12] Terns M.P. (2018). CRISPR-Based Technologies: Impact of RNA-Targeting Systems. Molecular cell, 72(3), 404-412. doi: 10.1016/j.molcel.2018.09.018

[13] Tang, L., Yang, F., He, X., Xie, H., Liu, X., Fu, J., et al. (2019). Efficient cleavage resolves PAM preferences of CRISPR-Cas in human cells. Cell regeneration (London, England), 8(2), 44-50. doi: 10.1016/j.cr.2019.08.002

[14] Munawar, N., & Ahmad, A. (2021). CRISPR/Cas System: An Introduction. CRISPR Crops: The Future of Food Security, 1-35. doi: 10.1007/978-981-15-7142-8_1

[15] Komor, A.C., Badran, A.H., & Liu, D.R. (2017). CRISPR-Based Technologies for the Manipulation of Eukaryotic Genomes. Cell, 168(1-2), 20-36. doi: 10.1016/j.cell.2016.10.044

[16] Sternberg, S.H., Redding, S., Jinek, M., Greene, E.C., & Doudna, J.A. (2014). DNA interrogation by the CRISPR RNA-guided endonuclease Cas9. Nature, 507(7490), 62-67. doi: 10.1038/nature13011

[17] Knott, G.J., & Doudna, J.A. (2018). CRISPR-Cas guides the future of genetic engineering. Science (New York, N.Y.), 361(6405), 866-869. doi: 10.1126/science.aat5011

[18] Hille, F., & Charpentier, E. (2016). CRISPR-Cas: biology, mechanisms and relevance. Philosophical transactions of the Royal Society of London. Series B, Biological sciences, 371(1707), 20150496. doi: 10.1098/rstb.2015.0496

[19] Zetsche, B., Gootenberg, J.S., Abudayyeh, O.O., Slaymaker, I.M., Makarova, K.S., Essletzbichler, P., et al. (2015). Cpf1 is a single RNA-guided endonuclease of a class 2 CRISPR-Cas system. Cell, 163(3), 759-771. doi: 10.1016/j.cell.2015.09.038

[20] Xu, X., & Qi, L.S. (2019). A CRISPR-dCas Toolbox for Genetic Engineering and Synthetic Biology. Journal of molecular biology, 431(1), 34-47. doi: 10.1016/j.jmb.2018.06.037

[21] Cello, J., Paul, A.V., & Wimmer, E. (2002). Chemical synthesis of poliovirus cDNA: generation of infectious virus in the absence of natural template. Science (New York, N.Y.), 297(5583), 1016-1018. doi:10.1126/science.1072266

[22] Smith, H.O., Hutchison, C.A., 3rd, Pfannkoch, C., & Venter, J.C. (2003). Generating a synthetic genome by whole genome assembly: ΦX174 bacteriophage from synthetic oligonucleotides. Proceedings of the National Academy of Sciences of the United States of America, 100(26), 15440-15445. doi: 10.1073/pnas.2237126100

[23] Gibson, D.G., Benders, G.A., Andrews-Pfannkoch, C., Denisova, E.A., Baden-Tillson, H., Zaveri, J., et al. (2008). Complete chemical synthesis, assembly, and cloning of a *Mycoplasma genitalium* genome. Science (New York, N.Y.), 319(5867), 1215-1220. doi: 10.1126/science.1151721

[24] Gibson, D.G., Glass, J.I., Lartigue, C., Noskov, V.N., Chuang, R.Y., Algire, M. A., et al. (2010). Creation of a bacterial cell controlled by a chemically synthesized genome. Science (New York, N.Y.), 329(5987), 52-56. doi: 10.1126/science.1190719

[25] Schindler D. (2020). Genetic Engineering and Synthetic Genomics in Yeast to Understand Life and Boost Biotechnology. Bioengineering (Basel, Switzerland), 7(4), 137. doi: 10.3390/bioengineering7040137

[26] Feyder, S., De Craene, J. O., Bär, S., Bertazzi, D. L., & Friant, S. (2015). Membrane trafficking in the yeast *S. cerevisiae* model. International journal of molecular sciences, 16(1), 1509-1525. doi: 10.3390/ijms16011509

[27] Pretorius, I.S., & Boeke, J.D. (2018). Yeast 2.0-connecting the dots in the construction of the world's first functional synthetic eukaryotic genome. FEMS yeast research, 18(4), foy032. doi: 10.1093/femsyr/foy032

[28] DiCarlo, J.E., Norville, J.E., Mali, P., Rios, X., Aach, J., & Church, G.M. (2013). Genome engineering in *S. cerevisiae* using CRISPR-Cas systems. Nucleic acids research, 41(7), 4336-4343. doi: 10.1093/nar/gkt135

[29] Farzadfard, F., Perli, S.D., & Lu, T.K. (2013). Tunable and multifunctional eukaryotic transcription factors based on CRISPR/Cas. ACS synthetic biology, 2(10), 604-613. doi: 10.1021/sb400081r

[30] Ryan, O.W., Skerker, J.M., Maurer, M.J., Li, X., Tsai, J.C., Poddar, S., et al. (2014). Selection of chromosomal DNA

libraries using a multiplex CRISPR system. eLife, 3, e03703. doi: 10.7554/eLife.03703

[31] Zhang, G.C., Kong, I.I., Kim, H., Liu, J.J., Cate, J.H., & Jin, Y.S. (2014). Construction of a quadruple auxotrophic mutant of an industrial polyploid strain by using RNA-guided Cas9 nuclease. Applied and environmental microbiology, 80(24), 7694-7701. doi:10.1128/AEM.02310-14

[32] Bao, Z., Xiao, H., Liang, J., Zhang, L., Xiong, X., Sun, N., Si, T., & Zhao, H. (2015). Homology-integrated CRISPR-Cas (HI-CRISPR) system for one-step multigene disruption in *S. cerevisiae*. ACS synthetic biology, 4(5), 585-594. doi: 10.1021/sb500255k

[33] Mans, R., van Rossum, H.M., Wijsman, M., Backx, A., Kuijpers, N.G., van den Broek, M., et al. (2015). CRISPR/Cas9: a molecular Swiss army knife for simultaneous introduction of multiple genetic modifications in *S. cerevisiae*. FEMS yeast research, 15(2), fov004. doi:10.1093/femsyr/fov004

[34] Ronda, C., Maury, J., Jakočiunas, T., Jacobsen, S.A., Germann, S.M., Harrison, S.J., et al. (2015). CrEdit: CRISPR mediated multi-loci gene integration in *S. cerevisiae*. Microbial cell factories, 14, 97. doi:10.1186/s12934-015-0288-3

[35] Jakočiūnas, T., Bonde, I., Herrgård, M., Harrison, S.J., Kristensen, M., Pedersen, L.E., et al. (2015). Multiplex metabolic pathway engineering using CRISPR/Cas9 in *S. cerevisiae*. Metabolic engineering, 28, 213-222. doi: 10.1016/j.ymben.2015.01.008

[36] Jakočiūnas, T., Rajkumar, A.S., Zhang, J., Arsovska, D., Rodriguez, A., Jendresen, C.B., et al. (2015). CasEMBLR: Cas9-Facilitated Multiloci Genomic Integration of in Vivo Assembled DNA Parts in *S. cerevisiae*. ACS synthetic biology, 4(11), 1226-1234. doi: 10.1021/acssynbio.5b00007

[37] Tsarmpopoulos, I., Gourgues, G., Blanchard, A., Vashee, S., Jores, J., Lartigue, C., et al. (2016). In-Yeast Engineering of a Bacterial Genome Using CRISPR/Cas9. ACS synthetic biology, 5(1), 104-109. doi:10.1021/acssynbio.5b00196

[38] Hao, H., Wang, X., Jia, H., Yu, M., Zhang, X., Tang, H., & Zhang, L. (2016). Large fragment deletion using a CRISPR/Cas9 system in *S. cerevisiae*. Analytical biochemistry, 509, 118-123. doi: 10.1016/j.ab.2016.07.008

[39] Chin, Y. W., Kang, W. K., Jang, H. W., Turner, T. L., & Kim, H. J. (2016). CAR1 deletion by CRISPR/Cas9 reduces formation of ethyl carbamate from ethanol fermentation by *S. cerevisiae*. Journal of industrial microbiology & biotechnology, 43(11), 1517-1525. doi: 10.1007/s10295-016-1831-x

[40] Biot-Pelletier, D., & Martin, V. J. (2016). Seamless site-directed mutagenesis of the *S. cerevisiae* genome using CRISPR-Cas9. Journal of biological engineering, 10, 6. doi: 10.1186/s13036-016-0028-1

[41] Shi, S., Liang, Y., Zhang, M. M., Ang, E. L., & Zhao, H. (2016). A highly efficient single-step, markerless strategy for multi-copy chromosomal integration of large biochemical pathways in *S. cerevisiae*. Metabolic engineering, 33, 19-27. doi: 10.1016/j.ymben.2015.10.011

[42] Sasano, Y., Nagasawa, K., Kaboli, S., Sugiyama, M., & Harashima, S. (2016). CRISPR-PCS: a powerful new approach to inducing multiple chromosome splitting in *S. cerevisiae*. Scientific reports, 6, 30278. doi: 10.1038/srep30278

[43] Jessop-Fabre, M. M., Jakočiūnas, T., Stovicek, V., Dai, Z., Jensen, M. K., Keasling, J. D., & Borodina, I. (2016). EasyClone-MarkerFree: A vector toolkit for marker-less integration of genes into *S. cerevisiae* via CRISPR-Cas9. Biotechnology journal, 11(8), 1110-1117. doi: 10.1002/biot.201600147

[44] Finnigan, G.C., & Thorner, J. (2016). mCAL: A New Approach for Versatile Multiplex Action of Cas9 Using One sgRNA and Loci Flanked by a Programmed Target Sequence. G3-Genes Genomes Genetics, 6(7), 2147-2156. doi: 10.1534/g3.116.029801

[45] Vanegas, K.G., Lehka, B.J., & Mortensen, U.H. (2017). SWITCH: a dynamic CRISPR tool for genome engineering and metabolic pathway control for cell factory construction in *S. cerevisiae*. Microbial cell factories, 16(1), 25. doi: 10.1186/s12934-017-0632-x

[46] Reider Apel, A., d'Espaux, L., Wehrs, M., Sachs, D., Li, R.A., Tong, G.J., et al. (2017). A Cas9-based toolkit to program gene expression in *S. cerevisiae*. Nucleic acids research, 45(1), 496-508. doi: 10.1093/nar/gkw1023

[47] Vigentini, I., Gebbia, M., Belotti, A., Foschino, R., & Roth, F.P. (2017). CRISPR/Cas9 System as a Valuable Genome Editing Tool for Wine Yeasts with Application to Decrease Urea Production. Frontiers in microbiology, 8, 2194. doi: 10.3389/fmicb.2017.02194

[48] Mans, R., Hassing, E.J., Wijsman, M., Giezekamp, A., Pronk, J.T., Daran, J.M., et al. (2017). A CRISPR/Cas9-based exploration into the elusive mechanism for lactate export in *S. cerevisiae*. FEMS yeast research, 17(8): 1-12. doi: 10.1093/femsyr/fox085

[49] Vanegas, K.G., Lehka, B.J., & Mortensen, U.H. (2017). SWITCH: a dynamic CRISPR tool for genome engineering and metabolic pathway control for cell factory construction in *S. cerevisiae*. Microbial cell factories, 16(1), 25. doi: 10.1186/s12934-017-0632-x

[50] Si, T., Chao, R., Min, Y., Wu, Y., Ren, W., & Zhao, H. (2017). Automated multiplex genome-scale engineering in yeast. Nature communications, 8, 15187. doi: 10.1038/ncomms15187

[51] Deaner, M., & Alper, H.S. (2017). Systematic testing of enzyme perturbation sensitivities via graded dCas9 modulation in *S. cerevisiae*. Metabolic engineering, 40, 14-22. doi: 10.1016/j.ymben.2017.01.012

[52] Deaner, M., Mejia, J., & Alper, H.S. (2017). Enabling Graded and Large-Scale Multiplex of Desired Genes Using a Dual-Mode dCas9 Activator in *S. cerevisiae*. ACS synthetic biology, 6(10), 1931-1943. doi: 10.1021/acssynbio.7b00163

[53] Jensen, E.D., Ferreira, R., Jakočiūnas, T., Arsovska, D., Zhang, J., Ding, L., et al. (2017). Transcriptional reprogramming in yeast using dCas9 and combinatorial gRNA strategies. Microbial cell factories, 16(1), 46. doi: 10.1186/s12934-017-0664-2

[54] Gander, M.W., Vrana, J.D., Voje, W.E., Carothers, J.M., & Klavins, E. (2017). Digital logic circuits in yeast with CRISPR-dCas9 NOR gates. Nature communications, 8, 15459. doi: 10.1038/ncomms15459

[55] Lian, J., HamediRad, M., Hu, S., & Zhao, H. (2017). Combinatorial metabolic engineering using an orthogonal tri-functional CRISPR system. Nature communications, 8(1), 1688. doi: 10.1038/s41467-017-01695-x

[56] Zhou, W., Tang, L., Cheng, K., Liu, M., Yang, Y., & Wang, W. (2017). [Construction of prototrophic glutathione-high-producing yeast strain mediated by CRISPR/Cas9]. Sheng wu gong cheng xue bao = Chinese journal of biotechnology, 33(12), 1999-2008. doi: 10.13345/j.cjb.170064

[57] Satomura, A., Nishioka, R., Mori, H., Sato, K., Kuroda, K., & Ueda, M. (2017). Precise genome-wide base editing by the CRISPR Nickase system in yeast. Scientific reports, 7(1), 2095. doi: 10.1038/s41598-017-02013-7

[58] Roggenkamp, E., Giersch, R.M., Wedeman, E., Eaton, M., Turnquist, E., Schrock, M.N., et al. (2017). CRISPR-UnLOCK: Multipurpose Cas9-Based Strategies for Conversion of Yeast Libraries and Strains. Frontiers in microbiology, 8, 1773. doi: 10.3389/fmicb.2017.01773

[59] Swiat, M.A., Dashko, S., den Ridder, M., Wijsman, M., van der Oost, J., Daran, J.M., et al. (2017). FnCpf1: a novel and efficient genome editing tool for *S. cerevisiae*. Nucleic acids research, 45(21), 12585-12598. doi: 10.1093/nar/gkx1007

[60] Ferreira, R., Skrekas, C., Nielsen, J., & David, F. (2018). Multiplexed CRISPR/Cas9 Genome Editing and Gene Regulation Using Csy4 in *S. cerevisiae*. ACS synthetic biology, 7(1), 10-15. doi:10.1021/acssynbio.7b00259

[61] Schultz, C., Lian, J., & Zhao, H. (2018). Metabolic Engineering of *S. cerevisiae* Using a Trifunctional CRISPR/Cas System for Simultaneous Gene Activation, Interference, and Deletion. Methods in enzymology, 608, 265-276. doi: 10.1016/bs.mie.2018.04.010

[62] Shao, Y., Lu, N., Qin, Z., & Xue, X. (2018). CRISPR-Cas9 Facilitated Multiple-Chromosome Fusion in *S. cerevisiae*. ACS synthetic biology, 7(11), 2706-2708. doi: 10.1021/acssynbio.8b00397

[63] Roy, K.R., Smith, J.D., Vonesch, S.C., Lin, G., Tu, C.S., Lederer, A.R., et al. (2018). Multiplexed precision genome editing with trackable genomic barcodes in yeast. Nature biotechnology, 36(6), 512-520. doi: 10.1038/nbt.4137

[64] Xue, T., Liu, K., Chen, D., Yuan, X., Fang, J., Yan, H., et al. (2018). Improved bioethanol production using CRISPR/Cas9 to disrupt the ADH2 gene in *S. cerevisiae*. World journal of microbiology & biotechnology, 34(10), 154. doi: 10.1007/s11274-018-2518-4

[65] Hou, S., Qin, Q., & Dai, J. (2018). Wicket: A Versatile Tool for the Integration and Optimization of Exogenous Pathways in *S. cerevisiae*. ACS synthetic biology, 7(3), 782-788. doi: 10.1021/acssynbio.7b00391

[66] Xie, Z.X., Mitchell, L.A., Liu, H.M., Li, B.Z., Liu, D., Agmon, N., et al. (2018). Rapid and Efficient CRISPR/Cas9-Based Mating-Type Switching of *S. cerevisiae*. G3 (Bethesda, Md.), 8(1), 173-183. doi: 10.1534/g3.117.300347

[67] Bao, Z., HamediRad, M., Xue, P., Xiao, H., Tasan, I., Chao, R., et al. (2018). Genome-scale engineering of *S. cerevisiae* with single-nucleotide precision. Nature biotechnology, 36(6), 505-508. doi:10.1038/nbt.4132

[68] Li, Z.H., Liu, M., Wang, F.Q., & Wei, D.Z. (2018). Cpf1-assisted efficient genomic integration of in vivo assembled DNA parts in *S. cerevisiae*. Biotechnology letters, 40(8), 1253-1261. doi:10.1007/s10529-018-2574-8

[69] Verwaal, R., Buiting-Wiessenhaan, N., Dalhuijsen, S., & Roubos, J. A. (2018). CRISPR/Cpf1 enables fast and simple genome editing of *S. cerevisiae*. Yeast (Chichester, England), 35(2), 201-211. doi: 10.1002/yea.3278

[70] Li, Z.H., Liu, M., Lyu, X.M., Wang, F.Q., & Wei, D.Z. (2018). CRISPR/Cpf1 facilitated large fragment deletion in *S. cerevisiae*. Journal of basic microbiology, 58(12), 1100-1104. doi: 10.1002/jobm.201800195

[71] Dank, A., Smid, E.J., & Notebaart, R.A. (2018). CRISPR-Cas genome engineering of esterase activity in *S. cerevisiae* steers aroma formation. BMC research notes, 11(1), 682. doi: 10.1186/s13104-018-3788-5

[72] Guo, X., Chavez, A., Tung, A., Chan, Y., Kaas, C., Yin, Y., et al. (2018). High-throughput creation and

functional profiling of DNA sequence variant libraries using CRISPR-Cas9 in yeast. Nature biotechnology, 36(6), 540-546. doi: 10.1038/nbt.4147

[73] Jakočiūnas, T., Pedersen, L.E., Lis, A.V., Jensen, M.K., & Keasling, J.D. (2018). CasPER, a method for directed evolution in genomic contexts using mutagenesis and CRISPR/Cas9. Metabolic engineering, 48, 288-296. doi: 10.1016/j.ymben.2018.07.001

[74] Soreanu, I., Hendler, A., Dahan, D., Dovrat, D., & Aharoni, A. (2018). Marker-free genetic manipulations in yeast using CRISPR/CAS9 system. Current genetics, 64(5), 1129-1139. doi: 10.1007/s00294-018-0831-y

[75] Zhang, Y., Wang, J., Wang, Z., Zhang, Y., Shi, S., Nielsen, J., et al. (2019). A gRNA-tRNA array for CRISPR-Cas9 based rapid multiplexed genome editing in *S. cerevisiae*. Nature communications, 10(1), 1053. doi: 10.1038/s41467-019-09005-3

[76] Laughery, M.F., Mayes, H.C., Pedroza, I.K., & Wyrick, J.J. (2019). R-loop formation by dCas9 is mutagenic in *S. cerevisiae*. Nucleic acids research, 47(5), 2389-2401. doi: 10.1093/nar/gky1278

[77] Shao, Y., Lu, N., Xue, X., & Qin, Z. (2019). Creating functional chromosome fusions in yeast with CRISPR-Cas9. Nature protocols, 14(8), 2521-2545. doi: 10.1038/s41596-019-0192-0

[78] Hofmann, A., Falk, J., Prangemeier, T., Happel, D., Köber, A., Christmann, A., et al. (2019). A tightly regulated and adjustable CRISPR-dCas9 based AND gate in yeast. Nucleic acids research, 47(1), 509-520. doi: 10.1093/nar/gky1191

[79] Ai, L., Guo, W., Chen, W., Teng, Y., & Bai, L. (2019). The gal80 Deletion by CRISPR-Cas9 in Engineered *S. cerevisiae* Produces Artemisinic Acid Without Galactose Induction. Current microbiology, 76(11), 1313-1319. doi: 10.1007/s00284-019-01752-2

[80] Chen, H., Zhu, C., Zhu, M., Xiong, J., Ma, H., Zhuo, M., et al. (2019). High production of valencene in *S. cerevisiae* through metabolic engineering. Microbial cell factories, 18(1), 195. doi: 10.1186/s12934-019-1246-2

[81] Laughery, M.F., & Wyrick, J.J. (2019). Simple CRISPR-Cas9 Genome Editing in *S. cerevisiae*. Current protocols in molecular biology, 129(1), e110. doi: 10.1002/cpmb.110

[82] Ciurkot, K., Vonk, B., Gorochowski, T.E., Roubos, J.A., & Verwaal, R. (2019). CRISPR/Cas12a Multiplex Genome Editing of *S. cerevisiae* and the Creation of Yeast Pixel Art. Journal of visualized experiments: JoVE, (147), 10.3791/59350. doi: 10.3791/59350

[83] Yan, Y., & Finnigan, G.C. (2019). Analysis of CRISPR gene drive design in budding yeast. Access microbiology, 1(9), e000059. doi: 10.1099/acmi.0.000059

[84] Cámara, E., Lenitz, I., & Nygård, Y. (2020). A CRISPR activation and interference toolkit for industrial *S. cerevisiae* strain KE6-12. Scientific reports, 10(1), 14605. doi: 10.1038/s41598-020-71648-w

[85] Levi, O., & Arava, Y. (2020). Expanding the CRISPR/Cas9 Toolbox for Gene Engineering in *S. cerevisiae*. Current microbiology, 77(3), 468-478. doi: 10.1007/s00284-019-01851-0

[86] Li, Z.H., Meng, H., Ma, B., Tao, X., Liu, M., Wang, F.Q., et al. (2020). Immediate, multiplexed and sequential genome engineering facilitated by CRISPR/Cas9 in *S. cerevisiae*. Journal of industrial microbiology & biotechnology, 47(1), 83-96. doi: 10.1007/s10295-019-02251-w

[87] Mitsui, R., Yamada, R., Matsumoto, T., Yoshihara, S., Tokumoto, H., & Ogino, H. (2020). Construction of lactic acid-tolerant *S. cerevisiae* by using CRISPR-Cas-mediated genome evolution for efficient D-lactic acid production. Applied microbiology and biotechnology, 104(21), 9147-9158. doi: 10.1007/s00253-020-10906-3

[88] Yang, S., Cao, X., Yu, W., Li, S., & Zhou, Y.J. (2020). Efficient targeted mutation of genomic essential genes in yeast *S. cerevisiae*. Applied microbiology and biotechnology, 104(7), 3037-3047. doi:10.1007/s00253-020-10405-5

[89] Xu, H., Han, M., Zhou, S., Li, B.Z., Wu, Y., et al. (2020). Chromosome drives via CRISPR-Cas9 in yeast. Nature communications, 11(1), 4344. doi: 10.1038/s41467-020-18222-0

[90] van Wyk, N., Kroukamp, H., Espinosa, M.I., von Wallbrunn, C., Wendland, J., & Pretorius, I.S. (2020). Blending wine yeast phenotypes with the aid of CRISPR DNA editing technologies. International journal of food microbiology, 324, 108615. doi: 10.1016/j.ijfoodmicro.2020.108615

[91] Gong, G., Zhang, Y., Wang, Z., Liu, L., Shi, S., Siewers, V., et al. (2021). GTR 2.0: gRNA-tRNA Array and Cas9-NG Based Genome Disruption and Single-Nucleotide Conversion in *S. cerevisiae*. ACS synthetic biology, 10(6), 1328-1337. doi: 10.1021/acssynbio.0c00560

[92] McGlincy, N.J., Meacham, Z.A., Reynaud, K.K., Muller, R., Baum, R., & Ingolia, N.T. (2021). A genome-scale CRISPR interference guide library enables comprehensive phenotypic profiling in yeast. BMC genomics, 22(1), 205. doi: 10.1186/s12864-021-07518-0

[93] Zhang, Z.X., Wang, Y.Z., Xu, Y.S., Sun, X.M., & Huang, H. (2021). Developing GDi-CRISPR System for Multi-copy Integration in *S. cerevisiae*. Applied biochemistry and biotechnology, 193(7), 2379-2388. doi: 10.1007/s12010-021-03532-w

[94] Heidari, R., Shaw, D.M., & Elger, B.S. (2017). CRISPR and the Rebirth of Synthetic Biology. Science and engineering ethics, 23(2), 351-363. doi: 10.1007/s11948-016-9768-z

[95] Cameron, D.E., Bashor, C.J., & Collins, J.J. (2014). A brief history of synthetic biology. Nature Reviews Microbiology, 12(5), 381-390.

[96] Sun, L., Lutz, B.M., & Tao, Y.X. (2016). The CRISPR/Cas9 system for gene editing and its potential application in pain research. Translational perioperative and pain medicine, 1(3), 22-33.

[97] Cho, S.W., Kim, S., Kim, Y., Kweon, J., Kim, H.S., Bae, S., et al. (2014). Analysis of off-target effects of CRISPR/Cas-derived RNA-guided endonucleases and nickases. Genome research, 24(1), 132-141. doi: 10.1101/gr.162339.113

[98] Doench, J.G., Fusi, N., Sullender, M., Hegde, M., Vaimberg, E.W., Donovan, K.F., et al. (2016). Optimized sgRNA design to maximize activity and minimize off-target effects of CRISPR/Cas9. Nature biotechnology, 34(2), 184-191. doi: 10.1038/nbt.3437

Chapter 4

CRISPR-Cas9: Role in Processing of Modular Metabolic Engineered Bio-Based Products

Vishnu Sharma, Tarun Kumar Kumawat, Anjali Pandit, Bhoomika Sharma, Pooja Agarwal, Bhawana Sharma, Preeti Nain and Manish Biyani

Abstract

Biogenetic engineering is a significant technology to sensibly manage microbial metabolic product factories. Genome modification methods for efficiently controlling and modifying genes at the genome level have progressed in biogenetic engineering during the last decade. CRISPR is genome editing technology that allows for the modification of organisms' genomes. CRISPR and its related RNA-guided endonuclease are versatile advanced immune system frameworks for defending against foreign DNA and RNAs. CRISPR is efficient, accessible, and trustworthy genomic modification tool in unparalleled resolution. At present, CRISPR-Cas9 method is expanded to industrially manipulate cells. Metabolically modified organisms are quickly becoming interested in the production of different bio-based components. Here, chapter explore about the control productivity of targeted biomolecules in divergent cells based on the use of different CRISPR-related Cas9.

Keywords: Biogenetic engineering, CRISPR, Endonuclease, Metabolic biomolecules

1. Introduction

The manufacture of biobased metabolic products by microbial production lines offers a viable path to a continuous future. At present, numerous bacterial strains have largely been employed to producediverse variety of metabolites that are useful for diverse industries including food and pharma [1, 2]. To increase the yield of metabolic products, genome editing is widely used. Genome modification is a form of genetic manipulation in which single bases of DNA are manipulated by adding, removing, or altering the genome of bacteria [3]. Despite it, most bacterial strains still face difficulties in genetic modification that is key impediment to metabolic engineering. Conventionally the zinc finger nucleases and transcription-activator like effector nucleases have been adopted for bacterial genetic modification [4, 5]. Both genetic modifications revolve around the principle of DNA-protein recognition [6].

ZFNs owned by SangamoBioSciences is one of the oldest gene-editing technologies established in the 1990s [7]. ZFNs are the engineered proteins that bind to the desired DNA. These proteins have two domains, the first one is a manufactured zinc-finger DNA binding domain and other is a DNA cleaving domain [8]. A basic zinc finger device has series of 4–6 binding modules. A codon is recognized by every unit [7]. Both domains are linked together via a chain of linker sequences. The DNA sequence of 24 bp is the first domain and other domain cleaves the recognized sequence in 5–7 bp spacer regions with the help of a restriction enzymeFokI [8]. FokI nucleases are type II's restriction enzymes that cause single-stranded breaks in a double-helical DNA strand. ZFN was withdrawn due to shortcomings such as the time-consuming and costly production of target enzymes, poor specificity, and elevated off-target variations, which were gradually overcome by the technological innovation [2, 7].

TALEN is another oldest gene-editing technology that was discovered as a replacement for ZFNs. It is made up of extremely repetitive DNA sequences that promote in-vivo homologous recombination. TALENs, like ZFNs, have two domains: N-terminal transcription activator-like effector (TALE) DNA-binding domain and C-terminal restriction endonuclease FokI catalytic domain [2, 8]. Both type of gene-editing are similar in having two sequence-specific DNA-binding proteins (two zinc-finger domains/TALEN domains) adjoining a target sequence, with the C-terminal of zinc-finger domain/TALEN domain being accompanied by a FokI enzyme, which cuts the target DNA in the form of a dimer [7, 9].

These methods, however, are hampered by the need to build a new nuclease pair for each genomic target. Both are also unable to target several genes at the same time. Therefore, due to complexity in designing, processing, and verifying the molecular requirements for nuclease expression and its targeting, both ZFNs and TALENs are escaped [10]. The CRISPR/Cas systems for genome editing are a novel technique that allows for the simultaneous targeting of numerous genes for the synthesis of superior strains.

2. CRISPR/Cas gene structure

The concept of CRISPR was introduced in 1987, whilst Japanese scientist Ishino and team were working on the *iap* gene in *Escherichia coli*. Entire gene encodes an alkaline phosphatase in *Escherichia coli*. They discovered repeated DNAs in bacterial genome that is not like other regular sequences [2, 11]. These re-occurring DNA sequences might be the components of recurring DNA sequences known as "Regularly clustered short palindrome repetitions" (CRISPR). Structurally, small repetitions of DNA are followed in CRISPR Systems by short spacer segments of genome, which are obtained via the standard bacterial path to a bacteriophage or plasmid. These repetitions are also related to nucleases or helicases in which particular DNA sequences are separated or unwinded [9, 12].

CRISPR is a bacterial and archaeal defense mechanism that works in hybrid with CRISPR-associated proteins. These were first discovered inside microorganisms DNA, but were subsequently extended to provide adaptive immune system for microorganisms [13]. CRISPR/Cas systems are easily adapted for genome modification because to their great practicality, comparative simplicity, and robustness [14]. CRISPR/Cas sequences are constituted of two or maybe more direct, often partially palindromatic, or frequently accurate repetitions (25–35 bp), which are separated into one or maybe more operon modules by single spacers (typically 30–40 bp) and an adjacent multiple-case cluster [15].

In genome editing, CRISPR works with a double-strand DNA cleavage at the particular target site near gRNA [16–18]. In process, CRISPR follows three separate but often interrelated stages: (i) acclimatization, (ii) pre-crRNA (pre-CRISPR RNA) expression and processing, and (iii) interference. The Cas protein complex

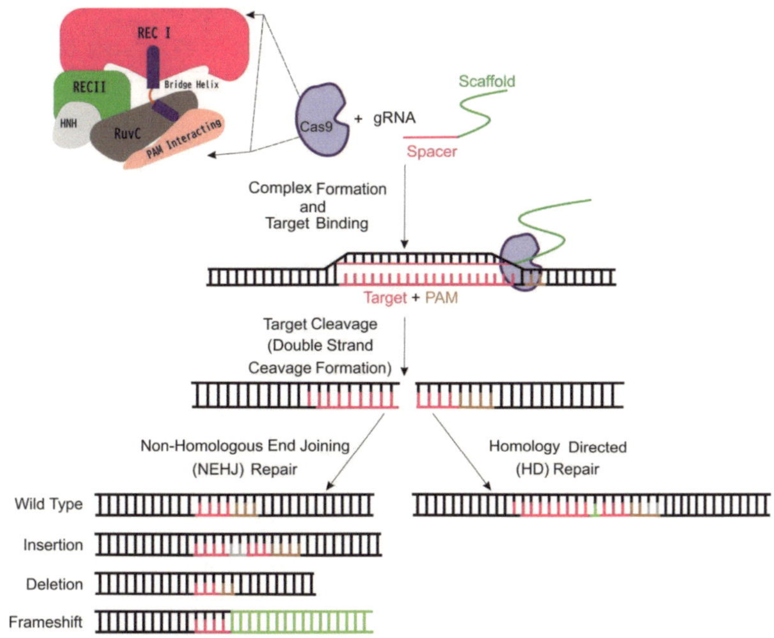

Figure 1.
Overview of CRISPR-Cas9 and recent developments in CRISPR/Cas9 genome editing.

attaches to a intended DNA molecule during the acclimatization stage and generally inserts two double-strand (ds) breakages into the target DNA, after a clear, short (2–4 bp) pattern, known as PAM(Protospacer-adjacent motif). The released fragment is subsequently transferred into the proximal repeat units of the CRISPR assortment. It is then fixed by cellular repair machinery, resulting in proximal repeat duplication [19–21].

Later, the CRISPR array is transcribed into a single long transcript by the expressive processing stage. The transcribed transcript is recognized as pre-crRNA, which is used for producing mature crRNAs using a distinct complex of proteins from Cas, a dedicated processing nuclease (Cas6), a single large Cas protein, or an external foundation. In the end, at interference stage, cRNA is utilized to detect protospacer that stay attached to the gRNA and then cleaved or inactivated by Cas nuclease [15]. The double-stranded cleavage partakes in DNA repair by essential cellular mechanisms. Usually, it entails non-homologous end-joining module and sometimes homology-directed repair [22, 23] (**Figure 1**). In between, the Cas9 is activated via forming single guide RNA molecule and triggers double-stranded cleavage at DNA target [24].

3. CRISPR classification

CRISPR-Cas systems display extraordinary diversity, including in core genes yielded by multiple CRISPR-Cas variations, gene structure, genomically locus architecture, and the original sequences [25–27]. The current CRISPR-Cas hierarchy contains three primary kinds (I, II and III), the less prevalent, but distinct, Type IV, V & VI on the basis of diversification of Cas genes (**Figure 2**) [28, 29]. Type I has the characteristic gene Cas3 that expresses the large protein with a helicase to unwind DNA–DNA and RNA–DNA duplexes. Sometimes the domain of helicases combines with an HD domain (conserved protein region with histidine (H) and/or

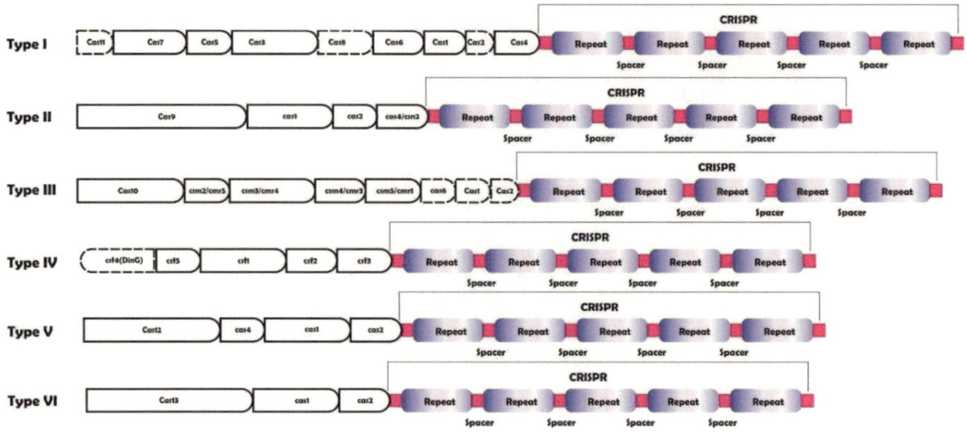

Figure 2.
The arrangement of several types of CRISPR-Cas systems.

aspartate (D) amino acid residues) and reveals endonuclease activity to make cleavage of the target DNA [28, 30, 31].

The type I Cas system contains the Cas1, Cas2, Cas5, Cas7, and Cas6 transcripts. Cas1 and Cas2 are genes encoding the cascade complex's components (which include big and small subunits). The Cas5, Cas7, and Cas6 loci are involved in the processing of pre-crRNA transcripts. Cas system type I is categorized into six sub-genotypes: I-A through I-F, each having its own unique gene and functioning organisms. In I-F complexes, Cas3 is also linked to Cas2 gene. In contrast to other variants, I-E and I-F are deficient in the Cas4 gene [32, 33].

The CRISPR-Type II contains the Cas9 gene, which codifies and controls the cascade complexes' functions through a multidomain protein [17]. Six domains make up Cas9 protein: REC I & II, Bridge Helix, PAM Associating, HNH, and RuvC [34, 35]. Rec-I is indeed the primary subunit responsible for RNA binding. The purpose of the REC II section is unknown presently. The arginine-rich coupled helix is the area that initiates cleavage when target DNA is bound [36]. The PAM-Interacting region aids in the definition of the PAM specificity necessary for target-DNA binding. The HNH and RuvC regions are nuclease areas that catalyze single-stranded DNA cleavage (**Figure 1**) [35]. The type-II CRISPR-Cas system has three sub - types: II-A, II-B, and II-C [37, 38]. Additionally, the type II-A system has the csn2 signature gene. Although this Csn2 gene has an unknown function, it produces tetrameric rings that interact with double-stranded Genetic material through center opening [39]. Because the type II-B Cas system lacks the csn2 gene, it retains a distinct Cas4 gene. The protein produced by this distinct gene functions as a 5′ DNA exonuclease [28]. Similarly, the type II-C Cas system contains the Cas1, Cas2, and Cas9 protein-coding genes. Cas type II have been widely embraced as a powerful tool for genomic editing [40].

The CRISPR-III systems possess Cas10 as main gene and encode a palm domain-like multi-domain protein related to that employed in PolBcyclases and polymerases. Cas10 is usually fused into an HD (histidine-aspartate) family nuclease region unique from CRISPR-Cas type I HD domains [41].

When encoding Cas1 and Cas2, CRISPR-III systems utilize crRNAs supplied by the Cas array linked to either a type I or type II Cas system. This system is classified into III-A through III-D subtypes. Csm, Cas1, Cas2, and Cas6 proteins are present in III-A type. Csm is a crRNA-guided enzyme that also acts as a DNase or occasionally as cyclic oligoadenylate kinase. Only Cmr proteins are present in the III-B pathway, which is lacking of Cas1, 2, and 6 loci. According to proximity sequence of crRNA,

Cmr identifies and degrades nucleic acids. A cyclase-inactivated Cas10 protein is discovered in the III-C type. Type III-C includes an inactivated cyclase domain Cas10 protein, while type III- includes an uncharacterized functional gene [42–44].

The CRISPR-Cas type-IV systems exist with plasmid genome of numerous bacteria. It lacks both Cas1 and Cas2, and not typically linked to CRISPR arrays andhas a high-decrease effector complex (CSF1). The CSF1 consist of csf1 (highly reduced subunit), Receptor activity-modifying protein encoding genes belonging to Cas5 (csf3) and Cas7 (csf2) family [28]. Although all CRISPR/Cas systems have certain functionality, Type II CRISPR/Cas is frequently adopted system that establishonly on Cas9 protein for the silencing of DNAi genes [45]. Cas9 protein is a large protein, involved in nucleic acid cleavage, with molecular weight of ~158 kDa. It has combine structure consisting of α-helical recognition and nuclease lobes [46].

The recognition lobe is made up of extended helix, REC1 and REC2 regions. Thenuclease region is generated of RuvC, HNH, and PAM-interacting (PI) C-terminal domain (CTD) [35, 47]. RuvC is named after the RuvC segment of *E. coli*, which decides formation of Holliday junctions [23]. In structure, protein motifs associates with spacer precursors or protospacers from the DNA of an attacking bacteriophage. These proto-spacer adjacent motifs are widely known as PAMs [9]. The crRNA and tracrRNA can be combined into guide-RNA, which enables the engagement of Cas9, which is necessary for double-stranded DNA cleavage [6].

The V-CRISPR-Cas12 system was designed for external genome editing applications such as gene expression suppression or activation, epigenome editing, in-situ genomic imaging, and large-scale genome screening [38, 47, 48]. CRISPR-Cas13 type VI is a tool for various RNA handling in the context of RNA interference (RNAi), in-vivo RNA visualization, and nucleic acid detection [49, 50].

4. Modern achievements in CRISPR-Cas9 mediated system

CRISPR/Cas9 technology has enabled a qualitative change in the range of gene functions for transcriptional control, gene targeting, epigenetic correction, gene therapy, and drug delivery of host genomes [51]. CRISPR/Cas technology possesses multiloci genome editing without the integration of a gene marker on the selection genome and saves time and exertion in metabolic engineering. Although several genetic modifications are available; the CRISPR/Cas9 technology significantly enhanced the efficiency of genetic engineering and is adopted as an extraordinary "gift." The CRISPR/Cas9 technology improved industrial micro-organisms' performance in strengthening of microbial factories that are valuable in processing of new value-added molecules from the low-cost feedstock.

There are abundant examples of bacteria, yeasts, and filamentous fungi which are reviewed in several studies of solicitations of the CRISPR/Cas9 system [51–57]. For example, *E. coli*, *S. cerevisiae*, *Bacillus sp.*, *Clostridium sp.*, *Corynebacterium sp.*, *Lactobacillus sp.*, *Mycobacterium sp.*, *Pseudomonas sp.*, *Streptomyces sp.* etc. [52, 58–69] are employed in the CRISPR/Cas system to improve yield of various metabolic products in field of industrial biotechnology. As a proof of concept, Zheng et al. employed Type I-F system to engineer *Zymomonas mobilis* as a synthetic chassis for sustainable economic biofuel and biochemical productions [70].

A study was designed to distinguish the orthogonal CRISPR method using *E. coli* for chromosomal addition of the Spd-Cas9based CRISPR module. Here found that out of SaCas9, St1Cas9, and FnCas12a, the St1Cas9 and SaCas9 are highly efficient to cause double stranded DNA break without associating with the sgRNA. This characteristic renders St1Cas9 into the *E. coli* chromosome as a hopeful Cas9 ortholog to combine whole or inadequate modules for succinate productionwith

Metabolic product	Engineering by CRISPR	Host organism	Outcome	Reference
2-Phenylethanol	Multiple genes cassette related to Shikimate pathway was targeted at the ABZ1 site with an efficiency of 51 ± 9%.	*Kluyveromyces marxianus*	The modified strain revealed the highest biosynthesis of 1943 ± 63 mg/L 2-phenylethanol.	[76]
2,3-Butanediol	Using CRISPR-Cas9, the gdh gene was targeted to produce (2R,3S)-BDO.	*Bacillus licheniformis*	As a consequence, fed-batch fermentation investigations showed stereospecific synthesis of (2R, 3S)-BDO.	[77]
5-Aminolevulinic Acid	The genes involved in TCA cycle were modified	*Shewanella oneidensis*	The downregulation of the essential hemB exhibited 2-fold increasing ALA production	[78]
Scleric Acid	The Cassette of crucial transcriptional repressor gene was activated to prevent the creation of an entirely new class of hybrid natural products.	*Streptomyces sclerotialus*	The biosynthetic route that encodes the synthesis of scleric acid.	[79]
β-Carotene	The β-carotene-rich cultivar was developed by targeting the fifth exon of the lycopene epsilon-cyclase (LCY) gene.	*Musa acuminata*	In comparison to wild genome, modified lines revealed a 6-fold increase in β-carotene concentration (~24 μg/g).	[80]
n-Butanol	Following the deletion of endogenous adhE gene into the efficient xylose-using host genome, a synthetic butanol pathway cassette was integrated.	*Escherichia coli*	At the bioreactor level, the modified strain produced 1.34 g/L butanol, which was 21-fold more than the parent strain.	[81]
2,3-Butanediol	In hostgenome, the 2,3-BDO biosynthesis pathway was introduced with presence of BDH1, alsS and alsDgenesfrom *Bacilus subtilis* and noxE gene from *Lactococcuslactis*.	*Saccharomyces cerevisiae*	Engineered strain produced remarkable amount (178 g/L) of 2,3-BDO from glucose instead of ethanol.	[82]
Itaconic Acid	Targeting of cyp3, MEL, UA and P_{ria1}, P_{etef} genes	*Ustilago maydis*	The deletion of by-product encoding genes enhanced itaconatetitre, rate, and yield.	[83]

Metabolic product	Engineering by CRISPR	Host organism	Outcome	Reference
Muconic Acid	The multiple genes (*CAN1*, RFP, TKL1, ARO4^{K229L}, ARO1$^{\Delta aroE}$, and ZWF1) were processed for upregulation and downregulation with a hybrid of CRISPR system and RNA interference.	*Saccharomyces cerevisiae*	The modified strain generated improved yield of cis,cis-muconic acid on feed-in-time medium.	[84]
Butyric Acid	Aconitase genes are suppressed in the synthetic butyrate pathway, and phosphotransferase and butyrate kinase genes are introduced.	*Corynebacterium glutamicum*	Altered strain revealed an improved yield of butyrate production (0.52 ± 0.02 g/L) than wild strain.	[85]
Octanoic Acid	Overexpression o fabZ and deletion of fade, fumAC and ackA genes	*Escherichia coli*	Product yield increased by 61% with a titer of 442 mg/l.	[86]
3-Hydroxybutyrate	Targeted to transcriptional repression of pta& aor2 genes	*Clostridium ljungdahlii*	Downregulating of pta gene increases the yield of 3-hydroxybutyrate with a 2.3-fold	[87]
Isopropanol	The gene cassettes thl, atoDA, adc, and adh or thl, ctfAB, adc, and adh were targeted in isopropanol synthetic pathway.	*Escherichia coli*	The modified strain produced maximum isopropanol productivity, above the original strain, of 0,62 g/l/h.	[88]
γ-Aminobutyric Acid	Three distinct genes (gabP, gabT, and Ncgl1221) were knockout to enhance the yield of product.	*Corynebacterium glutamicum*	The mutant strains expedite the production of γ-amino butyric acid metabolic products.	[89]
Galactaric Acid	The gene cassette encoding putative metabolic enzymes was removed.	*Aspergillus niger*	The modified strain generated galactaric acid from D-galacturonic acid. The modified strain was also able to convert pectin-rich biomass to galactaric acid.	[90]

Table 1.
CRISPR biotechnology applications in production of variable metabolic product.

178% improvement. It also efficiently hinders production of byproducts including lactate, formate, and ethanol [71].

Another research sought to increase CRISPR/Cas9 expression in methylotrophic fungus *Pichia pastoris*. Numerous genomic areas, including the Cas9 DNA sequence, gRNA regions, RNA synthetase II & III promoters, have been thoroughly examined and shown to have near-perfect targeting efficiency. Additionally, the altered strain was shown to be able to fulfill future requirements in synthetic biology, biotechnology, and metabolic pathway engineering. Zhang et al. focused on the soya bean plant's competing metabolic pathways for isoflavone production. Through the use of CRISPR/Cas9-mediated multiplex gene editing, the GmF3H1, GmF3H2, and GmFNSII-1 genes were deleted from the genistein competing route in this research [72].

Yang et al. utilized the RNP-based CRISPR–Cas9 technology to modify the genome of *Aspergillus niger* to increase succinic acid synthesis in CRISPR modified metabolic products. The desired strain was changed in this research by interrupting genes responsible to synthesize gluconic acid and oxalic acid. Indeed the C4-dicarboxylate transporter and the NADH-dependent fumarate reductase were overexpressed in this manner. The resultant strain generated 17 g/L succinic acid, while the wild-type strain grown on a synthetic substrate produced none [73].

Generally, genome modification in Schizosaccharomyces pombe is more complex than in *S. cerevisiae* owing to the reduced effectiveness of foreign DNA adjunction by homologous recombination [74]. As a result, Ozaki et al. modified the *S. pombe* strain using the CRISPR-Cas9 system and synthesized D-lactic acid from both glucose and cellobiose. The active genes for pyruvate decarboxylases, dehydrogenase, and glycerol-3-phosphate dehydrogenase were deleted in this research, and the D-lactate dehydrogenase gene from *Lactobacillus plantarum* was incorporated into the *S. pombegenome* [75]. The applications of CRISPR biotechnology to specified host species are outlined below in order to generate varied metabolic products. (**Table 1**).

5. Challenges in CRISPR/Cas9 applications

CRISPR/Cas9 provides tremendous genome-control capabilities, but there are still numerous obstacles to be overcome. The lack of a reliable DNA repair is the most significant of the difficulties associated with CRISPR/Cas technology, according to the researchers. As a consequence, numerous researches are increasing the CRISPR mechanisms, with the gene-editing technique likely to continual evolution for the foreseeable future. Similarly, lack of related techniques for creating single guide RNA is a distinct impediment. Limited methods for combining CRISPR/Cas9 with other genome-editing technologies, Cas9 endonuclease toxicity, off-target effects, the incidence of undesired mutations, and ethical issues are among the remaining issues. To counter these limitations, researchers have attempted to create and access various base editing approaches [91]. Besides, human genome has only one-sixteenth PAM sites, restrict the number of gene targetable sequences. So, novel Cas9 varieties are required to search and increase PAM interaction in the new experiments.

6. Future perspective

The future of new genetic mutations engineering should be to enhance the effectiveness of imminent models by joining innovative characteristics. In comparison to conventional genome editing systems, the CRISPR/Cas9 approach has provided rapid multiple genome sites editing of industrial strains at a time. Future models of CRISPR-Cas9 not only enable us to predict the success of editing but also the outcome. In this respect, the integration of droplet-based micro fluidics with

CRISPR/Cas9 could begin breakthroughs in modern biology. However, researchers can extract particular DNA segments but through micro homology can delete specific DNA segments and control CRISPR-Cas9 results. This approach enables to take advantage of the micro homology-mediated repair mechanism. These features will combine into both on- and off-target activity predictions for an optimal projected pipeline of CRISPR, where a Cas9 fusion protein will modify one target sequence into another without cleavage.

7. Conclusions

The CRISPR/Cas9 executes genome engineering technology feasible for utilization in many fields. The multiple genes targeting in a genome by CRISPR technology allows the learning of synergistic outcomes via the suppression of essential genes. Additionally, this approach sheds new light on design of many metabolite-producing microorganisms/bioreactors used in industrial biotechnology. However, certain drawbacks endure the potential uses of CRISPR-Cas systems. Conversely, the development of CRISPR-edited products and services faces sociopolitical obstacles, public acceptability, and government regulations. We must be stay update on the challenges by adding new features to improve CRISPR/Cas9 accuracy. We can anticipate that a lot of researchers from many fields concentrating their efforts towards this system will resolve the integrated limitations so that CRISPR will work its way into the emerging culture.

Author details

Vishnu Sharma[1*], Tarun Kumar Kumawat[1], Anjali Pandit[1], Bhoomika Sharma[1], Pooja Agarwal[1], Bhawana Sharma[1], Preeti Nain[1] and Manish Biyani[1,2]

1 Department of Biotechnology, Biyani Girls College, Jaipur, Rajasthan, India

2 Department of Bioscience and Biotechnology, Japan Advanced Institute of Science and Technology, Ishikawa, Japan

*Address all correspondence to: vishnusharma666@yahoo.com

References

[1] Lu H, Villada JC, Lee PKH. Modular Metabolic Engineering for Biobased Chemical Production. Trends in Biotechnology. 2019; 37(2):152-166.

[2] Fokum E, Zabed HM, Guo Q, Yun J, Yang M, Pang H, et al. Metabolic engineering of bacterial strains using CRISPR/Cas9 systems for biosynthesis of value-added products. Food Bioscience. 2019; 28: 125-32.

[3] Jacinto F V., Link W, Ferreira BI. CRISPR/Cas9-mediated genome editing: From basic research to translational medicine. Journal of Cellular and Molecular Medicine. 2020; 24(7):3766-3778.

[4] Gao W, Long L, Tian X, Xu F, Liu J, Singh PK, et al. Genome editing in cotton with the CRISPR/Cas9 system. Frontiers in Plant Science. 2017; 8:1364.

[5] Zhang Y, Massel K, Godwin ID, Gao C. Applications and potential of genome editing in crop improvement. Genome Biology. 2018; 19(1):210.

[6] Ran FA, Hsu PD, Wright J, Agarwala V, Scott DA, Zhang F. Genome engineering using the CRISPR-Cas9 system. Nature Protocols. 2013;8(11): 2281-2308.

[7] Arora L, Narula A. Gene editing and crop improvement using CRISPR-Cas9 system. Frontiers in Plant Science. 2017; 8:1932.

[8] Gupta D, Bhattacharjee O, Mandal D, Sen MK, Dey D, Dasgupta A, et al. CRISPR-Cas9 system: A new-fangled dawn in gene editing. Life Sciences. 2019; 232:116636.

[9] Bao A, Burritt DJ, Chen H, Zhou X, Cao D, Tran L-SP. The CRISPR/Cas9 system and its applications in crop genome editing. Critical Reviews in Biotechnology. 2019; 39(3):321-36.

[10] Doudna JA, Charpentier E. The new frontier of genome engineering with CRISPR-Cas9. Science. 2014;346 (6213):1258096.

[11] Ishino Y, Shinagawa H, Makino K, Amemura M, Nakatura A. Nucleotide sequence of the iap gene, responsible for alkaline phosphatase isoenzyme conversion in *Escherichia coli*, and identification of the gene product. Journal of Bacteriology. 1987; 169(12):5429-5433.

[12] Gong B, Shin M, Sun J, Jung CH, Bolt EL, Van Der Oost J, et al. Molecular insights into DNA interference by CRISPR-associated nuclease-helicase Cas3. Proceedings of the National Academy of Sciences of the United States of America. 2014; 111(46): 16359-16364.

[13] Khanzadi MN, Khan AA. CRISPR/Cas9: Nature's gift to prokaryotes and an auspicious tool in genome editing. Journal of Basic Microbiology. 2020; 60(2): 91-102.

[14] Kaur K, Gupta AK, Rajput A, Kumar M. Ge-CRISPR - An integrated pipeline for the prediction and analysis of sgRNAs genome editing efficiency for CRISPR/Cas system. Scientific Reports. 2016; 6:30870.

[15] Koonin E V., Makarova KS. Origins and evolution of CRISPR-Cas systems. Philosophical Transactions of the Royal Society B: Biological Sciences. 2019; 374(1772):20180087.

[16] Gasiunas G, Barrangou R, Horvath P, Siksnys V. Cas9-crRNA ribonucleoprotein complex mediates specific DNA cleavage for adaptive immunity in bacteria. Proceedings of the National Academy of Sciences of the United States of America. 2012; 109(39):E2579-E2586.

[17] Jinek M, Chylinski K, Fonfara I, Hauer M, Doudna JA, Charpentier E. A programmable dual-RNA-guided DNA endonuclease in adaptive bacterial immunity. Science. 2012; 337(6096): 816-821.

[18] Wang H, La Russa M, Qi LS. CRISPR/Cas9 in Genome Editing and beyond. Annual Review of Biochemistry. 2016; 85:227-264.

[19] Amitai G, Sorek R. CRISPR-Cas adaptation: Insights into the mechanism of action. Nature Reviews Microbiology. 2016; 14(2):67-76.

[20] Killelea T, Bolt EL. CRISPR-Cas adaptive immunity and the three Rs. Bioscience Reports. 2017;37: BSR20160297.

[21] Liu T, Liu Z, Ye Q, Pan S, Wang X, Li Y, et al. Coupling transcriptional activation of CRISPR-Cas system and DNA repair genes by Csa3a in *Sulfolobusislandicus*. Nucleic Acids Research. 2017;45(15):8978-8992.

[22] Doetschman T, Georgieva T. Gene Editing with CRISPR/Cas9 RNA-Directed Nuclease. Circulation Research. 2017; 120(5):879-894.

[23] Hsu PD, Lander ES, Zhang F. Development and applications of CRISPR-Cas9 for genome engineering. Cell. 2014; 157:1262-1278.

[24] Mougiakos I, Bosma EF, de Vos WM, van Kranenburg R, van der Oost J. Next Generation Prokaryotic Engineering: The CRISPR-Cas Toolkit. Trends in Biotechnology. 2016; 34(7):575-587.

[25] Koonin E V, Makarova KS, Zhang F. Diversity, classification and evolution of CRISPR-Cas systems. Current Opinion in Microbiology. 2017; 37: 67-78.

[26] Makarova KS, Wolf YI, Alkhnbashi OS, Costa F, Shah SA, Saunders SJ, et al. An updated evolutionary classification of CRISPR-Cas systems. Nature Reviews Microbiology. 2015; 13(11):722-736.

[27] Makarova KS, Wolf YI, Koonin E V. Classification and Nomenclature of CRISPR-Cas Systems: Where from Here? The CRISPR Journal. 2018; 1(5):325-336.

[28] Makarova KS, Koonin E V. Annotation and classification of CRISPR-Cas systems. Methods in Molecular Biology. 2015; 1311:47-75.

[29] Makarova KS, Wolf YI, Koonin E V. The basic building blocks and evolution of CRISPR-Cas systems. In: Biochemical Society Transactions. 2013; 41(6): 1392-1400.

[30] Mulepati S, Bailey S. Structural and biochemical analysis of nuclease domain of clustered regularly interspaced short palindromic repeat (CRISPR)-associated protein 3 (Cas3). Journal of Biological Chemistry. 2011; 286(36): 31896-31903.

[31] Sinkunas T, Gasiunas G, Fremaux C, Barrangou R, Horvath P, Siksnys V. Cas3 is a single-stranded DNA nuclease and ATP-dependent helicase in the CRISPR/Cas immune system. EMBO Journal. 2011; 30(7):1335-1342.

[32] Chakrabarti A, Desai P, Wickstrom E. Transposon Tn7 Protein TnsD Binding to *Escherichia coli* attTn7 DNA and Its Eukaryotic Orthologs. Biochemistry. 2004; 43(10):2941-2946.

[33] Kholodii GY, Mindlin SZ, Bass IA, Yurieva O V., Minakhina S V., Nikiforov VG. Four genes, two ends, and a res region are involved in transposition of Tn5053: a paradigm for a novel family of transposons carrying either a mer operon or an integron. Molecular Microbiology. 1995; 17(6):1189-1200.

[34] Jinek M, Jiang F, Taylor DW, Sternberg SH, Kaya E, Ma E, et al.

Structures of Cas9 endonucleases reveal RNA-mediated conformational activation. Science. 2014; 343(6176):1247997.

[35] Nishimasu H, Ran FA, Hsu PD, Konermann S, Shehata SI, Dohmae N, et al. Crystal structure of Cas9 in complex with guide RNA and target DNA. Cell. 2014; 156(5):935-949.

[36] Anders C, Niewoehner O, Duerst A, Jinek M. Structural basis of PAM-dependent target DNA recognition by the Cas9 endonuclease. Nature. 2014; 513(7519):569-573.

[37] Chylinski K, Makarova KS, Charpentier E, Koonin E V. Classification and evolution of type II CRISPR-Cas systems. Nucleic Acids Research. 2014; 42(10):6091-6105.

[38] Fonfara I, Richter H, BratoviÄ M, Le Rhun A, Charpentier E. The CRISPR-associated DNA-cleaving enzyme Cpf1 also processes precursor CRISPR RNA. Nature. 2016; 532(7600):517-521.

[39] Lee JW, Na D, Park JM, Lee J, Choi S, Lee SY. Systems metabolic engineering of microorganisms for natural and non-natural chemicals. Nature Chemical Biology. 2012; 8(6): 536-546.

[40] Pennisi E. The CRISPR craze. Science. 2013; 341: 833-836.

[41] Nickel L, Weidenbach K, Jağer D, Backofen R, Lange SJ, Heidrich N, et al. Two CRISPR-Cas systems in *Methano sarcinamazei* strain *Gö1* display common processing features despite belonging to different types I and III. RNA Biology. 2013; 10(5):779-791.

[42] Liu Z, Dong H, Cui Y, Cong L, Zhang D. Application of different types of CRISPR/Cas-based systems in bacteria., Microbial Cell Factories. 2020; 19(1): 172.

[43] Taylor DW, Zhu Y, Staals RHJ, Kornfeld JE, Shinkai A, Van Der Oost J, et al. Structures of the CRISPR-Cmr complex reveal mode of RNA target positioning. Science. 2015; 348(6234): 581-585.

[44] You L, Ma J, Wang J, Artamonova D, Wang M, Liu L, et al. Structure Studies of the CRISPR-Csm Complex Reveal Mechanism of Co-transcriptional Interference. Cell. 2019; 176(1-2): 239-253.

[45] Bhaya D, Davison M, Barrangou R. CRISPR-Cas systems in bacteria and archaea: Versatile small RNAs for adaptive defense and regulation. Annual Review of Genetics. 2011; 45(1):273-297.

[46] Raper AT, Stephenson AA, Suo Z. Functional Insights Revealed by the Kinetic Mechanism of CRISPR/Cas9. Journal of the American Chemical Society. 2018; 140(8): 2971-2984.

[47] Takei Y, Shah S, Harvey S, Qi LS, Cai L. Multiplexed Dynamic Imaging of Genomic Loci by Combined CRISPR Imaging and DNA Sequential FISH. Biophysical Journal. 2017; 112(9): 1773-1776.

[48] Zheng Y, Li J, Wang B, Han J, Hao Y, Wang S, et al. Endogenous Type I CRISPR-Cas: From Foreign DNA Defense to Prokaryotic Engineering. Frontiers in Bioengineering and Biotechnology. 2020; 8:62.

[49] Cox DBT, Gootenberg JS, Abudayyeh OO, Franklin B, Kellner MJ, Joung J, et al. RNA editing with CRISPR-Cas13. Science. 2017; 358(6366): 1019-1027.

[50] O'Connell MR. Molecular Mechanisms of RNA Targeting by Cas13-containing Type VI CRISPR–Cas Systems. Journal of Molecular Biology. 2019; 431:66-87.

[51] Khwatenge CN, Nahashon SN. Recent Advances in the Application of CRISPR/Cas9 Gene Editing System in Poultry Species. Frontiers in Genetics. 2021; 12:627714.

[52] Dicarlo JE, Norville JE, Mali P, Rios X, Aach J, Church GM. Genome engineering in *Saccharomyces cerevisiae* using CRISPR-Cas systems. Nucleic Acids Research. 2013; 41(7): 4336-4343.

[53] Jiang W, Bikard D, Cox D, Zhang F, Marraffini LA. RNA-guided editing of bacterial genomes using CRISPR-Cas systems. Nature Biotechnology. 2013; 31(3):233-239.

[54] Mougiakos I, Bosma EF, Ganguly J, van der Oost J, van Kranenburg R. Hijacking CRISPR-Cas for high-throughput bacterial metabolic engineering: advances and prospects. Current Opinion in Biotechnology. 2018; 50:146-157.

[55] Tarasava K, Liu R, Garst A, Gill RT. Combinatorial pathway engineering using type I-E CRISPR interference. Biotechnology and Bioengineering. 2018; 115(7):1878-1883.

[56] Wang Y, Wang S, Chen W, Song L, Zhang Y, Shen Z, et al. CRISPRCas9 and CRISPR-assisted cytidine deaminase enable precise and efficient genome editing in *Klebsiella pneumoniae*. Applied and Environmental Microbiology. 2018; 84(23):e01834-18.

[57] Yao R, Liu D, Jia X, Zheng Y, Liu W, Xiao Y. CRISPR-Cas9/Cas12a biotechnology and application in bacteria. Synthetic and Systems Biotechnology. 2018; 3(3):135-149.

[58] Yan Q, Fong SS. Challenges and advances for genetic engineering of non-model bacteria and uses in consolidated bioprocessing. Frontiers in Microbiology. 2017; 8, 2016.

[59] Choudhary E, Thakur P, Pareek M, Agarwal N. Gene silencing by CRISPR interference in mycobacteria. Nature Communications. 2015; 6(1):6267.

[60] Cobb RE, Wang Y, Zhao H. High-Efficiency Multiplex Genome Editing of *Streptomyces Species* Using an Engineered CRISPR/Cas System. ACS Synthetic Biology. 2015; 4(6):723-728.

[61] Jiang F, Doudna JA. CRISPR-Cas9 Structures and Mechanisms. Annual Review of Biophysics. 2017; 46:505-529.

[62] Joseph RC, Kim NM, Sandoval NR. Recent developments of the synthetic biology toolkit for Clostridium. Frontiers in Microbiology. 2018; 9:154.

[63] Li S, Jendresen CB, Grünberger A, Ronda C, Jensen SI, Noack S, et al. Enhanced protein and biochemical production using CRISPRi-based growth switches. Metabolic Engineering. 2016; 38:274-284.

[64] Lian J, Hamedirad M, Hu S, Zhao H. Combinatorial metabolic engineering using an orthogonal tri-functional CRISPR system. Nature Communications. 2017; 8(1):1688.

[65] Oh JH, Van Pijkeren JP. CRISPR-Cas9-assisted recombineering in *Lactobacillusreuteri*. Nucleic Acids Research. 2014; 42(17): e131.

[66] Tan SZ, Reisch CR, Prather KLJ. A robust CRISPR interference gene repression system in *Pseudomonas*. Journal of Bacteriology. 2018; 200(7): e00575-17.

[67] Tong Y, Whitford CM, Robertsen HL, Blin K, Jørgensen TS, Klitgaard AK, et al. Highly efficient DSB-free base editing for *streptomycetes* with CRISPR-BEST. Proceedings of the National Academy of Sciences of the United States of America. 2019; 116(41):20366-20375.

[68] Westbrook AW, Moo-Young M, Chou CP. Development of a CRISPR-Cas9 tool kit for comprehensive engineering of *Bacillus subtilis*. Applied and Environmental Microbiology. 2016; 82(16):4876-4895.

[69] Wang Y, Liu Y, Liu J, Guo Y, Fan L, Ni X, et al. MACBETH: Multiplex automated *Corynebacterium glutamicum* base editing method. Metabolic Engineering. 2018; 47:200-210.

[70] Zheng Y, Han J, Wang B, Hu X, Li R, Shen W, et al. Characterization and repurposing of the endogenous Type I-F CRISPR-Cas system of *Zymomonasmobilis* for genome engineering. Nucleic acids research. 2019; 47(21):11461-11475.

[71] Sung LY, Wu MY, Lin MW, Hsu MN, Truong VA, Shen CC, et al. Combining orthogonal CRISPR and CRISPRi systems for genome engineering and metabolic pathway modulation in *Escherichia coli*. Biotechnology and Bioengineering. 2019; 116(5):1066-1079.

[72] Zhang P, Du H, Wang J, Pu Y, Yang C, Yan R, et al. Multiplex CRISPR/Cas9-mediated metabolic engineering increases soya bean isoflavone content and resistance to soya bean mosaic virus. Plant Biotechnology Journal. 2020; 18(6):1384-1395.

[73] Oizumi Y, Kaji T, Tashiro S, Takeshita Y, Date Y, Kanoh J. Complete sequences of *Schizosaccharomyces pombe* sub *telomeres* reveal multiple patterns of genome variation. Nature Communications. 2021; 12(1):611.

[74] Ozaki A, Konishi R, Otomo C, Kishida M, Takayama S, Matsumoto T, et al. Metabolic engineering of *Schizosaccharomyces pombe* via CRISPR-Cas9 genome editing for lactic acid production from glucose and cellobiose. Metabolic Engineering Communications. 2017; 5:60-67.

[75] Yang L, Henriksen MM, Hansen RS, Lübeck M, Vang J, Andersen JE, et al. Metabolic engineering of *Aspergillus niger* via ribonucleoprotein-based CRISPR–Cas9 system for succinic acid production from renewable biomass. Biotechnology for Biofuels. 2020; 13(1):206.

[76] Li M, Lang X, Moran Cabrera M, De Keyser S, Sun X, Da Silva N, et al. CRISPR-mediated multigene integration enables Shikimate pathway refactoring for enhanced 2-phenylethanol biosynthesis in *Kluyveromyces marxianus*. Biotechnology for Biofuels. 2021; 14(1):3.

[77] Song CW, Rathnasingh C, Park JM, Kwon M, Song H. CRISPR-Cas9 mediated engineering of *Bacillus licheniformis* for industrial production of (2R,3S)-butanediol. Biotechnology Progress. 2021; 37(1): e3072.

[78] Yi YC, Ng IS. Redirection of metabolic flux in *Shewanella oneidensis* MR-1 by CRISPRi and modular design for 5-aminolevulinic acid production. Bioresources and Bioprocessing. 2021; 8(1):13.

[79] Alberti F, Leng DJ, Wilkening I, Song L, Tosin M, Corre C. Triggering the expression of a silent gene cluster from genetically intractable bacteria results in scleric acid discovery. Chemical Science. 2019; 10(2):453-463.

[80] Kaur N, Alok A, Shivani, Kumar P, Kaur N, Awasthi P, et al. CRISPR/Cas9 directed editing of lycopene epsilon-cyclase modulates metabolic flux for β-carotene biosynthesis in banana fruit. Metabolic Engineering. 2020; 59:76-86.

[81] Abdelaal AS, Jawed K, Yazdani SS. CRISPR/Cas9-mediated engineering of *Escherichia coli* for n-butanol production from xylose in defined medium. Journal of Industrial Microbiology and Biotechnology. 2019; 46(7): 965-975.

[82] Lee YG, Seo JH. Production of 2,3-butanediol from glucose and cassava hydrolysates by metabolically engineered industrial polyploid *Saccharomycescerevisiae*. Biotechnology for Biofuels. 2019; 12(1): 204.

[83] Becker J, Hosseinpour Tehrani H, Gauert M, Mampel J, Blank LM, Wierckx N. An *Ustilagomaydis* chassis for itaconic acid production without by-products. Microbial Biotechnology. 2020; 13(2):350-362.

[84] Kildegaard KR, Tramontin LRR, Chekina K, Li M, Goedecke TJ, Kristensen M, et al. CRISPR/Cas9-RNA interference system for combinatorial metabolic engineering of *Saccharomyces cerevisiae*. Yeast. 2019; 36(5):237-247.

[85] Yoon J, Woo HM. CRISPR interference–mediated metabolic engineering of *Corynebacterium glutamicum* for homo-butyrate production. Biotechnology and Bioengineering. 2018; 115(8):2067-2074.

[86] Tan Z, Yoon JM, Chowdhury A, Burdick K, Jarboe LR, Maranas CD, et al. Engineering of E. coli inherent fatty acid biosynthesis capacity to increase octanoic acid production. Biotechnology for Biofuels. 2018; 11(1):87.

[87] Woolston BM, Emerson DF, Currie DH, Stephanopoulos G. Rediverting carbon flux in *Clostridium ljungdahlii* using CRISPR interference (CRISPRi). Metabolic Engineering. 2018; 48:243-253.

[88] Liang L, Liu R, Garst AD, Lee T, Nogué VS i., Beckham GT, et al. CRISPR EnAbled Trackable Genome Engineering for isopropanol production in *Escherichia coli*. Metabolic Engineering. 2017; 41:1-10.

[89] Cho JS, Choi KR, Prabowo CPS, Shin JH, Yang D, Jang J, et al. CRISPR/Cas9-coupled recombineering for metabolic engineering of *Corynebacteriumglutamicum*. Metabolic Engineering. 2017; 42:157-167.

[90] Kuivanen J, Wang YMJ, Richard P. Engineering *Aspergillus niger* for galactaric acid production: Elimination of galactaric acid catabolism by using RNA sequencing and CRISPR/Cas9. Microbial Cell Factories. 2016; 15(1):210.

[91] Uddin F, Rudin CM, Sen T. CRISPR Gene Therapy: Applications, Limitations, and Implications for the Future. Frontiers in Oncology. 2020; 10:1387.

Chapter 5

Synthetic Gene Circuits for Antimicrobial Resistance and Cancer Research

Kevin S. Farquhar, Michael Tyler Guinn, Gábor Balázsi and Daniel A. Charlebois

Abstract

Mathematical models and synthetic gene circuits are powerful tools to develop novel treatments for patients with drug-resistant infections and cancers. Mathematical modeling guides the rational design of synthetic gene circuits. These systems are then assembled into unified constructs from existing and/or modified genetic components from a range of organisms. In this chapter, we describe modeling tools for the design and characterization of chemical- and light-inducible synthetic gene circuits in different organisms and highlight how synthetic gene circuits are advancing biomedical research. Specifically, we demonstrate how these quantitative model systems are being used to study drug resistance in microbes and to probe the spatial–temporal dimensions of cancer in mammalian cells.

Keywords: antimicrobial resistance, synthetic gene circuits, mathematical models, optogenetics, cancer

1. Introduction

A primary goal of synthetic biology is to rationally design and engineer synthetic gene circuits as tools to advance basic research [1, 2], optimize the production of chemicals or biofuels [3, 4], build biocomputational systems [5], and enhance clinical therapeutics [6]. Control of synthetic gene circuits at the transcriptional level (transcription is the process of transcribing mRNA from a DNA template) has been demonstrated through chemical- and light-based stimuli [7, 8]. The transcriptional network architecture (how genes are connected to and regulate each other through transcription factor proteins) affects the properties of gene expression, in terms of average expression levels as well as the degree of expression variability inside a single cell or across a cell population [9]. Throughout this chapter, we will use the term "synthetic gene circuits" to describe synthetic systems and the term "gene networks" to described natural systems.

Fluctuations in the biochemical processes of transcription and translation (translation is the process of translating amino acid-based proteins from a nucleotide-based mRNA template) are referred to as gene expression noise [10]. Gene expression noise leads to heterogeneity among genetically identical cells in the same environment and can affect the survival of microorganisms [11]. For instance, gene expression noise has been shown to promote drug resistance in microbes [12]. Similarly, gene expression noise is thought to play an important role in tumorigenesis and the development of resistance during cancer chemotherapy [13]. Mathematical models and synthetic gene circuits have established that the architecture of the gene network modulates gene expression noise [14].

Electronic circuits inspired the development of synthetic gene circuits, with mathematical representations of natural and synthetic networks successfully predicting their effects on gene expression [15]. A milestone study in *E. coli* demonstrated that negative feedback stabilizes the gene circuit's response to expression fluctuations [16]. The first synthetic toggle switch circuit in *E. coli* mimicked the electronic version and served as a simplified version of the naturally occurring bacteriophage lambda switch [17, 18]. The construction of a synthetic biological clock in *E. coli* permitted oscillations in gene expression to be tuned to a particular frequency [19]. By mimicking natural gene networks, synthetic gene circuits generate insights on how complex biological systems work by breaking down natural networks into their components, which is highly beneficial in basic biomedical research [20].

Optogenetics is the control of cellular components using electromagnetic radiation. Like other synthetic systems, optogenetic components can be engineered into gene circuits to precisely control cellular processes such as gene expression or protein activity in prokaryotic and eukaryotic cells; the performance of optogenetic gene circuits can be optimized in an iterative model-experiment cycle. However, unlike previous gene circuits, optogenetics offers the ability to control gene expression at a single-cell resolution. The fast temporal and single-cell spatial resolutions that light provides as a stimulus for gene circuits is unmatched; chemical stimulus regulates transcription on longer timescales and at a cell-population level. Like their gene circuit predecessors, optogenetic gene circuits can be used to control functional proteins. Optogenetic tools are especially suited to investigate gene function at the single-cell level. For instance, researchers can take a gene of interest, such as KRAS which is often found mutated in cancers [21], and integrate it into an optogenetic gene circuit to explore the transcriptional and translational effects on cellular phenotypes by stimulating individual cells with visible light. It is worth noting that although in this chapter we focus on optogenetic applications involving visible light, some optogenetic tools have been developed using other regions of the electromagnetic spectrum, including near-infrared [22] and UV [23] radiation.

This chapter describes the construction and characterization of synthetic gene circuits in yeast and mammalian cells (Section 2) and optogenetic gene circuits in mammalian cells (Section 3) with various transcriptional network architectures, along with their applications in biomedical research. The mathematical approaches to model synthetic and optogenetic gene circuits are also discussed.

2. Synthetic gene circuits

2.1 Positive feedback gene circuits in yeast

A positive feedback synthetic gene circuit was first constructed in yeast to convert a continuous gradient of a constitutively expressed transcriptional activator into a cell phenotype switch, resembling analog to digital signal conversion [24]. Subsequently, a positive feedback (PF) gene circuit was genomically integrated into

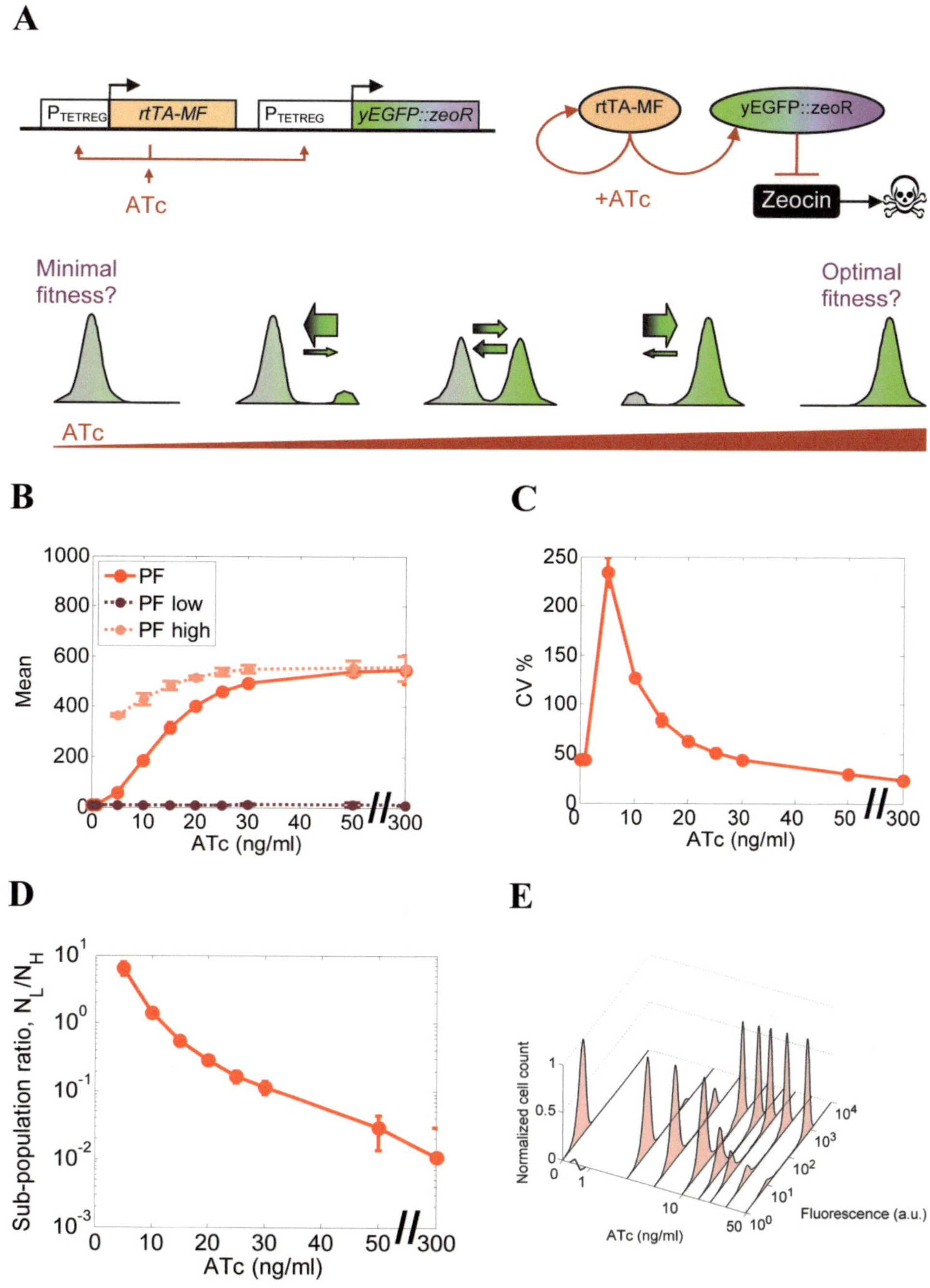

Figure 1.
(A) Schematic of the positive feedback (PF) synthetic gene circuit in yeast (top left). The regulator rtTA is toxic when active, but the gene circuit prevents Zeocin toxicity by activating the ZeoR gene (top right). The role of cellular memory in optimizing fitness is shown in the bottom panel. (B-E) dose responses. (B) Mean expression levels of the PF gene circuit and of the cell sorted low- and high-expression states. (C) Gene expression noise as determined from the coefficient of variation (CV). (D) Subpopulation ratio of low-expressing cells to high-expressing cells. (E) Steady-state gene expression distributions at a single-cell resolution. From ***Figure 2*** *"characterization of gene expression in cells bearing PF circuit" by Nevozhay, D. and Adams, R. et al. in [25] located at https://journals.plos.org/ploscompbiol/article?id=10.1371/journal.pcbi.1002480 under a CC BY 4.0 license with panel label font modified.*

the budding yeast *Saccharomyces cerevisiae* to investigate how cell population fitness (growth rate) and subpopulations emerge from the molecular-level kinetics of gene networks and single-cell division rates [25]. Synthetic gene circuits with positive feedback and cooperativity can display bistability, where cells switch between two gene expression states with a cellular "memory" that corresponds to the temporal maintenance of each state [26, 27].

In the PF gene circuit, the regulator reverse tetracycline-controlled trans-activator (rtTA) binds to its own promoter in the presence of tetracyclines (**Figure 1A**, top left) [28]. The genetic engineering approaches mirrored the assembly of a negative feedback (NF) circuit in yeast [29]. Unlike the NF circuit, toxicity exists after activating the regulator gene rtTA, which sequesters general transcription factors from vital cellular processes [30]. Additionally, the construct controlled the drug resistance gene ZeoR, which confers resistance to the antibiotic Zeocin (**Figure 1A**, top right).

The PF gene circuit exhibits a sigmoidal gene expression dose response at the population level (**Figure 1B**). The expression dose–responses for low- and high-expressing subpopulations were determined using a bimodality detection algorithm [28]. The gene expression noise level peaked at an intermediate inducer (anhydrotetracycline or ATc) concentration (**Figure 1C**). The shift in the gene expression distribution peaks over increasing inducer levels is reflected by the subpopulation ratio changes (**Figure 1D**). The gene expression distributions for the ATc dose response display bimodality (**Figure 1E**).

Testing multiple levels of inducer and drug determined a "fitness landscape" that quantitatively mapped population growth rates to unique combinations of ATc and Zeocin concentrations. The yeast PF system demonstrated the need to incorporate the cellular memory associated with gene expression states (**Figure 1A**, bottom) to computationally predict the fitness landscape [31]; the growth rate under Zeocin treatment was the highest at the minimal level of induction that lead to bimodal expression [28]. These computational predictions guided laboratory experiments to identify environmental conditions that defined a "sweet spot" of drug resistance, which balanced the costs of expressing rtTA with benefits of expressing ZeoR.

The yeast PF synthetic gene circuit was subsequently used to study evolutionary dynamics under various levels of induction and drug treatment [32]. This microbial evolution experiment included conditions at one edge or both edges (saturating molecular levels or none) of the fitness landscape, as well as at intermediate levels of induction with or without Zeocin. The evolved populations were frozen at specific time points and subsequently reestablished to test in various conditions compared to ancestral cells. Additionally, the evolution experiments were combined with simulations to predict the types of mutations that could arise under induction and drug treatment [32].

Full induction without drug treatment led to breakdown of the regulator rtTA over time through full and partial knock-out mutations, which improved fitness without Zeocin [32]. Yet, during follow-up evolution experiments in high induction with Zeocin partial rtTA knockout mutants regained function [31]. Full Zeocin treatment without induction eventually established populations with higher expression, potentially through mutations in the drug resistance gene and promoter linked with extra-circuit mutations. High induction with Zeocin led to accumulated mutations in rtTA, possibly lowering the effectiveness of the regulator as shown by decreased inducer sensitivity [32]. After intermediate induction with Zeocin, experiments on evolved populations during reintroduction of drug under intermediate induction uncovered the two expression distribution peaks shifting towards each other leading to a single fitness peak; this highlighted the role of noise in driving evolution through the trade-off between rtTA toxicity and drug resistance.

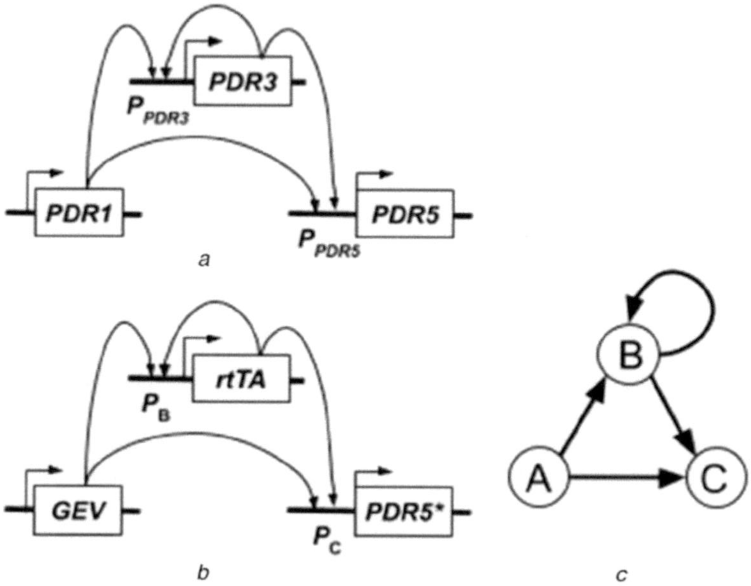

Figure 2.
(a) Natural pleiotropic drug resistance (PDR) network in yeast. (b) Synthetic PDR gene circuit. (c) Core PDR network architecture. From ***Figure 1*** *"network schematics" by Brendan Camellato et al. in [27] located at https://ietresearch.onlinelibrary.wiley.com/doi/10.1049/enb.2019.0009 under a CC BY-NC-ND 3.0 license.*

2.2 Experiments and computational models of positive feedback and feedforward circuits in yeast

Along with positive feedback and negative feedback circuit architectures, feedforward loop (FFL) architectures (a three-gene network composed of two input transcription factors, one of which regulates the other, both jointly regulating a target gene) may have evolved in natural gene regulatory networks to enhance fitness [26, 33]. In *S. cerevisiae*, the pleiotropic drug resistance (PDR) network contains a positive feedback loop embedded in a feedforward loop (FFL + PF) (**Figure 2a**) [34]. The PDR network provides multidrug resistance through an ABC transporter pump protein encoded by the PDR5 gene. A similar FFL + PF network may enhance drug resistance in human cancer cells [35].

Mathematical models of gene regulatory networks can predict biological responses, which is essential to optimally design synthetic gene circuits and to guide experiments. A minimal model of the PDR network found that the positive feedback and feedforward loop architectures sustain transcription and can stabilize expression of the network when the drug is transient or fluctuating [33]. The minimal model also predicted increased gene expression noise (in terms of increased noise magnitude and longer cellular memory timescales) in the FFL and FFL + PF networks. Overall, the FFL and FFL + PF network architectures were found to enhance drug resistance *in silico*.

The minimal model of the PDR network was described by the following system of coupled ordinary differential equations (ODEs) [33]:

$$\frac{dPDR3}{dt} = \alpha_{PDR3}\omega_1 f_{PDR3}(PDR1, PDR3) - PDR3 \tag{1}$$

$$\frac{dPDR5}{dt} = \alpha_{PDR5} f_{PDR5}(PDR1, PDR3) - PDR5$$

where PDR1 was treated as an adjustable parameter. α_{PDR3} and α_{PDR5} are the maximum levels of activated protein production for the variables PDR3 and PDR5, respectively. The Boolean parameter ω_1 describes the activation of PDR3 by PDR1. Here the dilution and degradation rates of PDR3 and PDR5 were set to unity. The functions that describe how PDR3 and PDR5 are regulated are given by:

$$
\begin{aligned}
f_{PDR3}(PDR1, PDR3) &= \frac{(PDR1 + \omega_2 PDR3)^n}{K^n + (PDR1 + \omega_2 PDR3)^n} \\
f_{PDR5}(PDR1, PDR3) &= \frac{(PDR1 + PDR3)^n}{K^n + (PDR1 + PDR3)^n}
\end{aligned}
\tag{2}
$$

where, n and K are the Hill coefficient and half-maximal activation parameter, respectively. The Boolean parameter ω_2 describes the presence or absence of positive feedback regulation on PDR3. The minimal model and a more comprehensive model (presented below) were translated into biochemical reactions that were simulated using the Gillespie stochastic simulation algorithm [36, 37].

A more comprehensive model, known as the PDR5 transcriptional network model, incorporated the dynamics of the PDR5 efflux protein pump and the negative feedback produced when PDR5 eliminates the drug from the cell [33]. The PDR5 transcriptional network model can be described by the following system of coupled ODEs:

$$
\begin{aligned}
\frac{dPDR1}{dt} &= \alpha_0 + \alpha_{PDR1} \frac{D_{int}}{K_{PDR1} + D_{int}} - \delta_{PDR1} PDR1 \\
\frac{dPDR3}{dt} &= \alpha_{PDR3} \frac{(PDR1 + PDR3)^{n_{PDR3}}}{K_{PDR3}^{n_{PDR3}} + (PDR1 + PDR3)^{n_{PDR3}}} - \delta_{PDR3} PDR3 \\
\frac{dPDR5}{dt} &= \alpha_{PDR5} \frac{(PDR1 + PDR3)^{n_{PDR5}}}{K_{PDR5}^{n_{PDR5}} + (PDR1 + PDR3)^{n_{PDR5}}} - \delta_{PDR5} PDR5 \\
\frac{dD_{int}}{dt} &= k_{diff}(D_{ext} - D_{int}) - k_{pump} PDR5 \left(\frac{D_{int}}{K_{int} + D_{int}} \right)
\end{aligned}
\tag{3}
$$

where PDR1 and the intracellular drug concentration (D_{int}) were incorporated as variables. α_0 is the basal rate of PDR1 transcription, D_{ext} the extracellular drug concentration, k_{diff} the rate of passive diffusion of the drug across the cellular membrane, k_{int} half-maximum saturation coefficient for the PDR5 efflux pump, and k_{pump} the efflux rate of the drug via PDR5 efflux pump. It was assumed that drug entry and exit from the cells occurred through a combination of passive and active transport, and that the activation of PDR1 by the drug can be described by Michaelis–Menten kinetics. The PDR5 transcriptional network model predicted that PDR5 expression level would increase after application of the drug and that cell population fitness would oscillate before stabilizing during drug treatment.

To confirm the predictions from the PDR5 transcriptional network model, a synthetic gene circuit (**Figure 2b**) was constructed with molecular cloning techniques [38] and integrated into a yeast strain without a native PDR5 gene [27]. The construction technique ran multiple overlap PCR steps, where two fragments with overlapping regions on their ends were amplified together initially without primers, using the overlapping sequence as a *de facto* primer. Homologous recombination facilitated the integration of the synthetic gene circuit into the yeast genome. This synthetic PDR gene circuit has rtTA activating its own expression through a tetracycline-inducible promoter (**Figure 2b**), mimicking the positive feedback activation of PDR3 (**Figure 2a**). The Doxycycline inducible promoter regulating rtTA expression is also controlled by a β-estradiol inducible GEV regulator, representing PDR1, which also activates the PDR5 gene in this synthetic gene circuit [27], completing the core PDR network architecture (**Figure 2c**) [34].

The PDR synthetic gene circuit was experimentally compared to simplified PDR circuit components, like direct activation (DA: PDR1 directly activates PDR5), cascade (CAS: PDR1 activates PDR5 through PDR3), a cascade with positive feedback loop (CAS + PF: PDR1 activates PDR5 through PDR3, with PDR3 activating its own expression), and a feedforward loop (FFL: PDR1 activates PDR5 directly as well as indirectly through PDR3) [27]. The FFL + PF circuit represented the PDR network. This separation of components tested the effect of specific network motifs in drug resistance and gene expression compared to the full PDR synthetic circuit. Direct activation in the FFL circuit was found experimentally to increase the speed of expression changes compared to indirect activation of PDR5. Direct activation was required for PDR5-mediated drug resistance. Indirect activation enhanced drug resistance, which provided evidence that the delayed reduction in PDR5 expression protected cells from the toxic effects of drug (Cycloheximide) exposure (**Figure 3a**). **Figure 3b** shows that strains carrying a gene circuit with a positive feedback loop grew faster in the presence of drug compared to strains in which this circuit architecture was missing. This provided evidence that the positive feedback regulation of PDR5 expression protects cells from drug exposure. Finally, reducing the strength of positive feedback regulation through mutation decreased drug resistance (**Figure 3c**).

Overall, the PDR network was recapitulated in mathematical and synthetic gene circuit models that demonstrated that the network architecture is optimized for drug resistance, with gene expression noise making important contributions to fitness during drug treatment.

2.3 Experiments and computational models of mammalian negative and positive feedback gene circuits

The yeast NF gene circuit was transferred to mammalian cells to test an organism transfer workflow using design and optimization cycles aided by computational modeling [40]. This workflow led to updates to the yeast NF gene circuit, including the optimization of the tet promoter operator site locations, the introduction of an intron upstream of the regulator, codon optimization for mammalian translation,

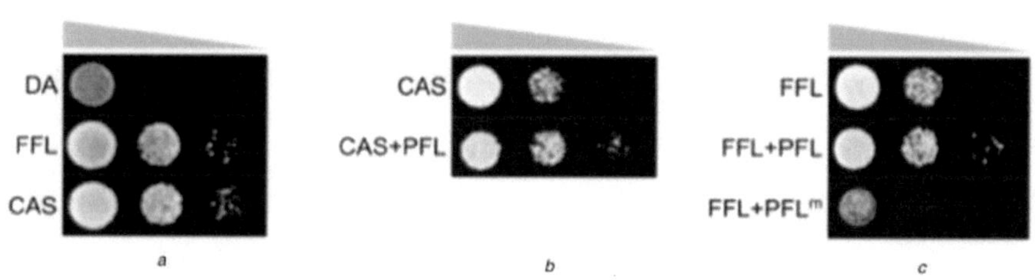

Figure 3.
*(a) Genetically engineered yeast strains with direct activation (DA) and indirect activation networks (FFL and CAS). (b) Yeast strains with indirect activation alone (CAS) or combined with positive feedback (CAS + PFL). (c) Yeast strains with feedforward activation alone (FFL) or combined with positive feedback (FFL +PFL) or mutated positive feedback (FFL+PFLm). Plates contain 5 μg/ml doxycycline (inducer), 0.025 (a) or 0.05 μg/ml (b, c) Cycloheximide (drug) and no β-estradiol (inducer). Gray triangles represent decreasing density of the spotted cell culture, 1:10 serial dilutions from OD600 0.1 to 1 x 10–3. From **Figure 6** "indirect activation and positive feedback enhance drug resistance in spot assay experiment" by Brendan Camellato et al. in [27] located at https://ietresearch.onlinelibrary.wiley.com/doi/10.1049/enb.2019.0009 under a CC BY-NC-ND 3.0 license.*

and the addition of a Kozak sequence near the start codon, which was stably but randomly transfected into MCF-7 breast cancer cells. This NF circuit in mammalian cells exhibited a linear dose response with low gene expression noise, similar to the NF circuit in yeast cells [29]. Though the adaptability of the yeast NF circuit to mammalian cells did not require any additional design features, optimization of parts responsible for gene expression and protein location was required to replicate the features of the yeast NF circuit. These results support the "abstraction principle" in the field of synthetic biology, namely that different parts of a biological network can be optimized for improved functionality in new settings, while leaving the original network design intact [40].

In a subsequent study, the yeast PF circuit was transferred to mammalian cells, which was coupled with a Flp-recombinase site-specific integration system (**Figure 4a**) [39]. This mammalian positive feedback (mPF) circuit displayed a sigmoidal mean gene expression dose response (**Figure 4b**). Gene expression noise increased at intermediate inducer (Doxycycline) levels (**Figure 4c**), with broad unimodal gene expression distributions (**Figure 4d**). A lack of bimodality is unexpected for an induced bistable circuit and may have been attributed to similar cellular growth and cellular memory time scales.

The previously reported mammalian negative feedback circuit was also connected to the Flp-In integration system in Chinese Hamster Ovary (CHO) cells (**Figure 5a**) [40]. This negative feedback circuit was separately integrated into the same genomic site as the mPF circuit and subsequently called the mammalian negative feedback (mNF) circuit. The mNF circuit displayed a linear dose response in mean gene expression (**Figure 5b**). Gene expression noise was low across all inducer (Doxycycline) levels (**Figure 5c**) with narrow gene expression distributions (**Figure 5d**).

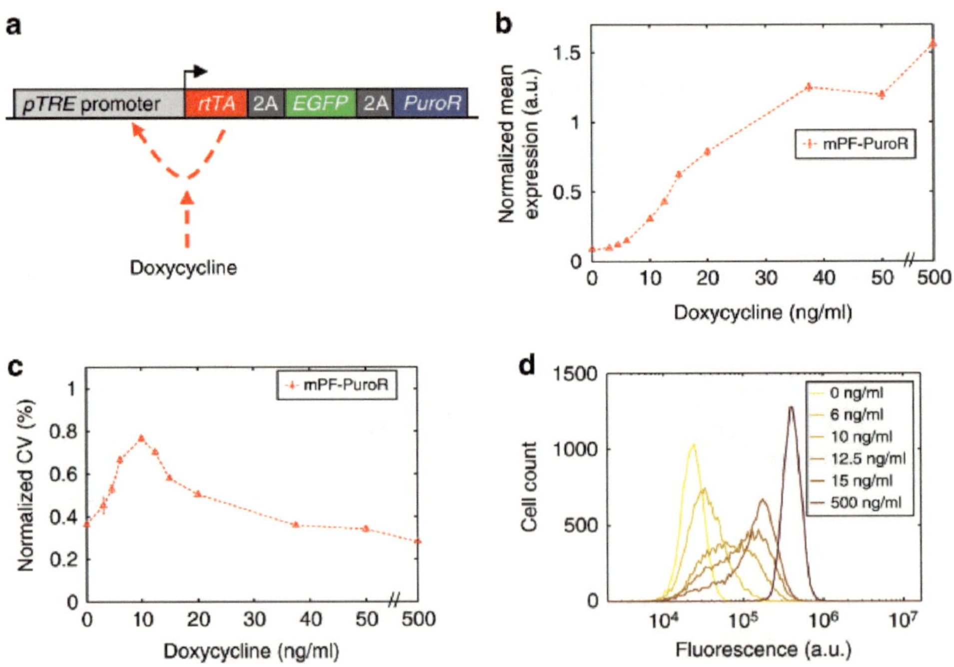

Figure 4.
*(a) Schematic of the mammalian positive feedback (mPF) synthetic gene network. (b-d) dose responses. (b) Mean gene expression. (c) Gene expression noise determined from the coefficient of variation (CV). (d) Single-cell gene expression distributions. From **Figure 2** "dose–response of the mPF-PuroR gene circuit" by Farquhar, K.S. et al. in [39] located at https://www.nature.com/articles/s41467-019-10330-w under a CC BY 4.0 license.*

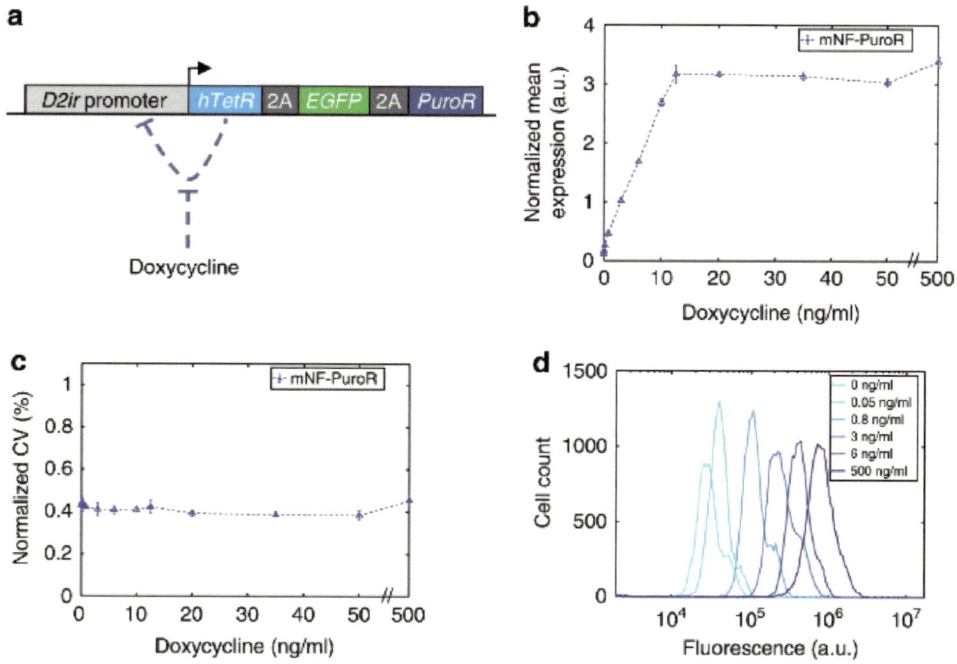

Figure 5.
*(a) Schematic of the mammalian negative feedback (mNF) synthetic gene network. (b-d) dose responses. (b) Mean gene expression. (c) Gene expression noise determined from the coefficient of variation (CV). (d) Single-cell gene expression distributions. From **Figure 3** "dose–response of the mNF-PuroR gene circuit" by Farquhar, K.S. et al. in [39] located at https://www.nature.com/articles/s41467-019-10330-w under a CC BY 4.0 license.*

The mPF and mNF gene circuits controlled the EGFP fluorescent protein and the PuroR drug resistance gene each separated by self-cleaving 2A motifs [39, 40]. The, the integration into the same genomic site and the introduction of self-cleaving 2A motifs and the PuroR drug resistance gene did not affect the function of these circuits.

In an evolution experiment with multiple drug (Puromycin) concentrations, the mNF and mPF circuits were tuned to the same mean expression level to decouple gene expression noise from the mean gene expression prior to drug treatment [39]. After adaptation, the drug was removed while induction was either maintained or removed. Finally, the adapted populations were retreated with the previous level of drug to uncover potential adaptation mechanisms. The evolution experiment demonstrated that low gene expression noise from the mNF circuit was beneficial in adaptation compared to mPF under low levels of drug [39]. In contrast, the high noise from the mPF circuit was beneficial compared to the mNF circuit under high levels of drug.

Mutations were found in the TetR regulator gene from the mNF circuit that knocked out repression, which explained why the mNF populations maintained high expression with or without inducer after the temporary removal of the drug. The mPF circuits did not mutate, which was consistent with the drug retreatment period where uninduced mPF populations struggled to adapt while the induced populations adapted faster [39]. Overall, the decoupling of gene expression noise from the mean demonstrated the power of using synthetic gene circuits to uncover novel insights into mammalian drug resistance.

A stochastic population dynamics model was developed to predict the emergence and switching dynamics of persister (P), nongenetically drug-resistant (N), and genetically drug-resistant (G) subpopulations (**Figure 6a**), which was described mathematically by the following set of coupled ODEs [39]:

$$\begin{aligned} \frac{dP}{dt} &= r_{P,N}N - r_{N,P}P - r_{G,P}P \\ \frac{dN}{dt} &= -r_{P,N}N + r_{N,P}P - r_{G,N}N + k_N N - g_N N \\ \frac{dG}{dt} &= r_{G,P}P + r_{G,N}N + k_G G - g_G G \end{aligned} \tag{4}$$

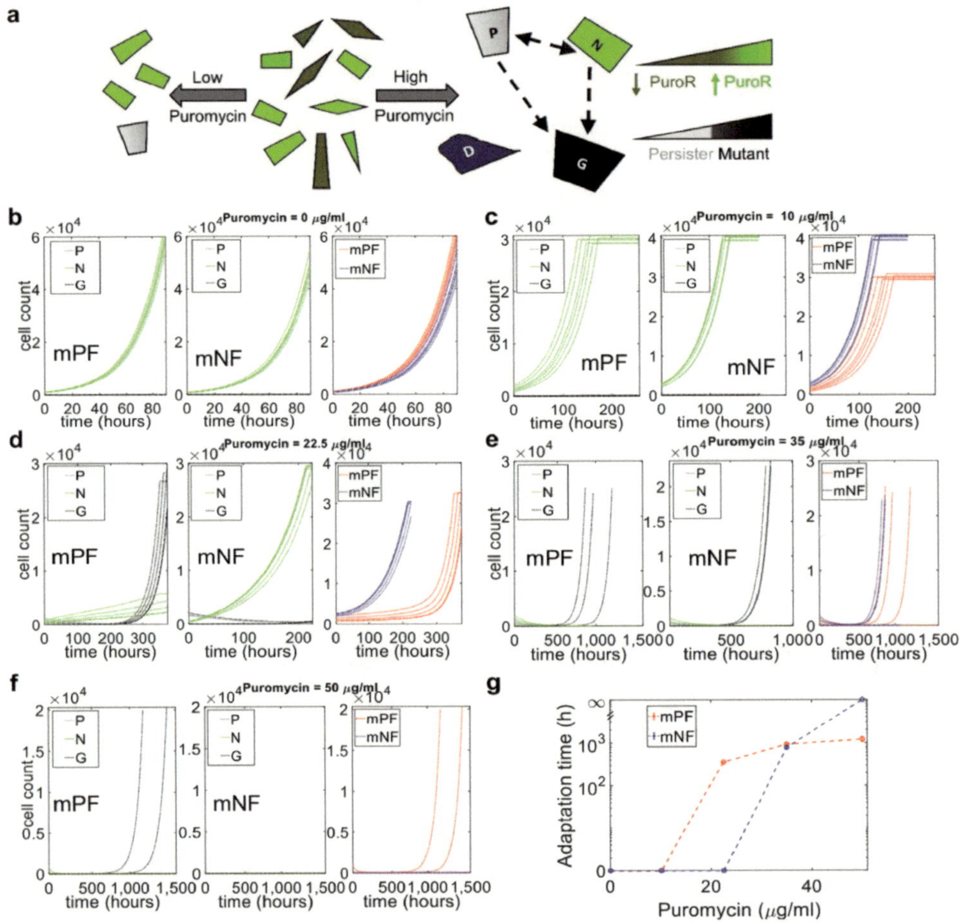

Figure 6.
(a) Schematic depicting the effects of drug (Puromycin) concentration on Chinese hamster ovary (CHO) cell population composition and survival. Nongenetically drug-resistant cells (green cells – Brighter cells have higher PuroR expression level and are therefore more resistant) and nongrowing persister cells (gray cells) can switch phenotypes (dashed bidirectional arrow). Persister cells and growing nongenetically resistant cells can also become stably drug-resistant cells (black cells). When no drug is present, a genetically identical (clonal) cell population with heterogeneous gene expression exists (center). Under low drug treatment conditions (left arrow), cells with low PuroR expression perish and a small fraction of the surviving clonal cells become persister cells. For high drug treatment conditions (right arrow), only cells with high PuroR expression levels can survive drug treatment while the rest die (dark blue cells), and a higher fraction of the surviving cells become persisters. As persister and nongenetically resistant cells can become stably drug resistant, the population on the right panel becomes increasingly heterogeneous over the course of treatment. (b-f) Representative growth curves for simulated mPF-PuroR and mNF-PuroR CHO cell populations under (b) 0, (c) 10, (d) 22.5, (e) 35, and (f) 50 μg/mL of Puromycin. Growth curves shown in panels in (b-f) correspond to: (left) mPF subpopulations, (center) mNF subpopulations, and (right) mPF and mNF populations. (g) Adaptation times corresponding to the mPF-PuroR and mNF-PuroR populations shown in panels (b-f). From ***Figure 6*** *"Modeling the adaptation of mPF-PuroR and mNF-PuroR cells in various concentrations of Puromycin" by Farquhar, K.S. et al. in [39] located at https://www.nature.com/articles/s41467-019-10330-w under a CC BY 4.0 license.*

where $r_{i,j}$ is transition rate from genotype or phenotype j to i, k_i is the growth rate of i, and g_i is the death rate of i. Noise was incorporated into this model by drawing the parameters describing the initial number of cells that survived Puromycin treatment and the carrying capacity of the cell culture environment from a normal distribution. Numerical simulations of Eq. (4) are shown in **Figure 6b–f**. These simulation results agreed with the data from the evolution experiments [39]. The modeling indicated that nongenetic phenotypic variability could facilitate the adaptation of the mPF and mNF strains to lower drug concentrations (**Figure 6c**), but that population dynamics in terms of the P to G conversion was required to capture the long experimental adaptation times at higher drug concentrations (**Figure 6d–g**).

3. Optogenetic gene circuits

A major focus of synthetic biology has been to engineer gene circuits to control cellular processes. This has mainly been achieved through small molecules that activate or inactivate various components of synthetic gene circuits [17, 19, 41–46]. Chemical stimuli has many advantages, including easy titration for inducing gene expression over large dynamic ranges, characterized affinity for existing proteins, and minimal off-target effects [9]. However, controlling gene circuits with chemicals is often not instantaneous and makes it difficult to control individual cells in a population.

Light stimulus can achieve many of the same advantages as chemicals without the above limitations. Like the cellular proteins that respond to chemical stimuli, proteins have been found in nature that respond to light [47, 48]. The discovery of light-activated proteins provided the elements necessary to build optogenetic gene circuits. By engineering light-responsive elements with existing components in gene circuits, synthetic biologists were able to adapt endogenous proteins from natural organisms to experimental model organisms, including yeast [7] and mammalian cell lines [49].

The use of light-inducible systems in eukaryotic organisms has expanded to cover nearly as many applications as chemical systems, including the control of gene expression, protein alterations, metabolic reactions, epigenetic states, and animal behavior [50–55]. A common theme among light-activated and chemical-induced circuits is the genetic architecture of the system. For chemically regulated gene circuits, classic engineering architectures [26] have been produced including negative regulation, positive regulation, positive feedback, negative feedback, and many others [39, 40, 42, 56–59]. Optogenetic systems have begun to incorporate these circuit architectures [49, 60]. Optogenetic tools respond to a variety of wavelengths of light [61, 62] and can be used transiently [63, 64] or as stable systems [60].

Negative feedback is an important gene circuit architecture that has been implemented in optogenetic circuits. Negative feedback is a desirable architecture in synthetic biology because it provides two advantages: 1) negative feedback reduces gene expression noise and 2) negative feedback allows tunability of system output to a "transfer function", which describes the relationship between an input and an output function [65, 66]. In synthetic biology, many systems are designed with desired inputs and outputs in mind and therefore knowing the relationship or transfer function between these variables is crucial [67]. Additionally, such features also occur in natural systems [68, 69], which synthetic systems are often designed to mimic [14]. The negative feedback circuit architecture has been engineered into synthetic gene circuits in bacterial, fungal, and mammalian systems, all controllable by small chemical molecules [29, 40, 70, 71]. This circuit architecture was recently engineered in an optogenetic system and found to offer many of the same advantages as the chemical-induced negative feedback gene circuits; namely, low gene circuit noise, wide system tunability, and a characterized transfer function between input and output (i.e., light and a fluorescence reporter) [60].

The optogenetic NF system (**Figure 7A**) was inspired from previous chemical gene circuits [40] and from computational modeling [57]. A well-known tetracycline-responsive system [72, 73] provided the foundation to engineer a light-responsive system, by fusing the TetR protein with a LOV2 domain [52, 74, 75] and either a degradation tag [74] or a small peptide [76] that inhibits TetR (**Figure 7B**). When light is absent, the degradation tag or the inhibitory peptide remains hidden. When blue light is present, the LOV2 protein undergoes a confirmational change and reveals one of the two domains. By employing this engineered light protein, an optogenetic gene circuit can be constructed with operator sites upstream of the gene for this protein to allow down regulation of its own expression (as well as another functional gene). The light stimulus can then be used to control gene expression output with the benefits of low noise and titratable expression levels.

Computational modeling was used to investigate how system performance could be enhanced in the optogenetic NF system [60]. This methodology of build, model, improve is crucial when developing synthetic gene circuits. To achieve this, the design and construction of the optogenetic NF system focused on changes that could decrease gene circuit noise, lower basal expression of the circuits, increase fold-change of the circuit, and enhance the range of circuit response to stimuli. A

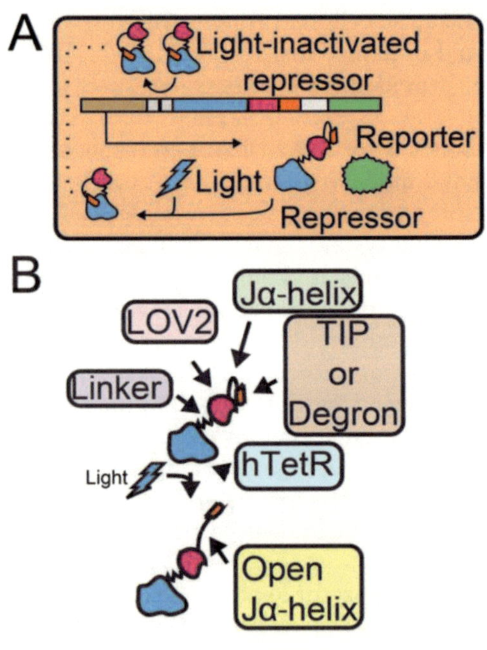

Figure 7.
(A) Schematic illustrating an optogenetic gene circuit with a negative feedback architecture. Specifically, this architecture produces a transcription factor (blue) that inhibits its own production. This transcription factor is also engineered to have a light-responsive domain (pink) and inhibitory peptide (orange). When light is added to the system, a confirmational change occurs and the hidden inhibitory peptide is exposed to inhibit the DNA-binding function of the transcription factor. When this occurs, increased transcription and translation occurs for the reporter gene. (B) Schematic illustration of the transcription factor shown in (A). The transcription factor is the TetR protein fused with a linker peptide which is fused with the light-responsive LOV2 domain, which contains a Jα-helix that is fused with a functional domain such as a degradation tag or an inhibitory peptide (TIP). When light is added, the Jα-helix opens exposing the functional domain. Figure used with permission from Guinn [55].

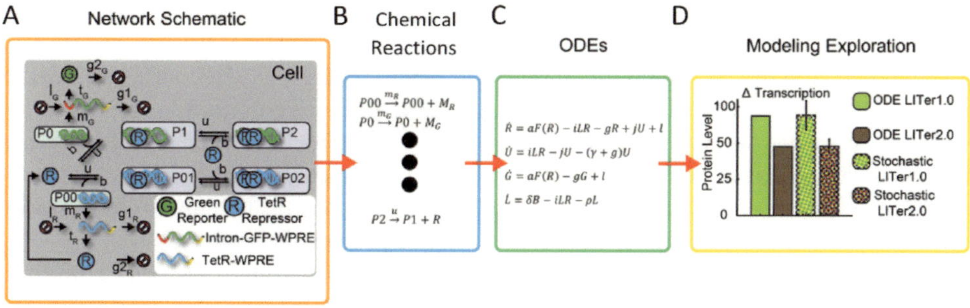

Figure 8.
The "pipeline" used to model the NF optogenetic gene circuit. The network schematic of the system of interest includes transcription factors, promoters that can be bound by transcription factors, reporters, RNA, etc. (A). The network schematic is converted into a set of chemical reactions (B). These chemical reactions are then converted into a set of ordinary differential equations (ODEs) (C). Finally, the ODEs are simulated numerically, providing quantitative predictions that can be used to improve the optogenetic gene circuit (D). Figure used with permission from Guinn [55].

quantitative gene expression model identified parameters that optimized the performance of the optogenetic NF system. The "pipeline" for modeling the NF optogenetic gene circuit is shown in **Figure 8**. The NF optogenetic circuit was represented by a network schematic of the main optogenetic gene circuit's components (**Figure 8A**). These components were formalized as a set of chemical reactions (**Figure 8B**). The chemical reactions were then described by a system of ordinary differential equations (**Figure 8C**), which was solved numerically [57]. Lastly, the equations describing each reaction were explored to investigate whether a given optogenetic gene circuit component should be changed experimentally (**Figure 8D**). The design of the optogenetic NF system was improved by changing the transcription and translation rates of the optogenetic inhibitory protein.

In addition to validation and improving optogenetic gene circuits, various architectures can be utilized for expressing functional proteins at precise levels inside of single cells. Controlling gene expression in single cells can allow for exploration of phenotypic landscapes as a function of protein levels and time. For example, the optogenetic NF system was used to control the mutated oncogene KRAS (G12V), which showed expression and function could be controlled in a dose-responsive manner with low optogenetic gene circuit noise [60]. This system can be modified to contain any functional gene allowing single-cell gene expression studies using microscopy equipment such as digital mirror device (DMD) [77]. DMD technology can allow system feedback for controlling optogenetic gene circuits *in silico* at the computer-microscopy interface [78] and *in vitro* using gene architecture designs responsive to light. Coupling technology like the DMD with optogenetic gene circuits like the optogenetic NF system will allow researchers to better understand cellular processes and single-cell biology.

Overall, optogenetic gene circuits allow researchers to perturb single cells to distinguish between individual and population-level behavior. Optogenetic tools are anticipated to be important for elucidating mechanisms in drug resistance and cancer metastasis, where single-cell behavior and spatial–temporal factors may dictate biological fate.

4. Conclusions

The ability of synthetic gene circuits to fulfill engineered design principles and facilitate scientific discoveries is expected to grow over time. However, evolutionary forces can undermine the integrity of synthetic gene circuits [25, 32]. It will be crucial to design gene circuits in the future to mitigate the effects of evolution to maintain their functional integrity. One approach is to use DNA sponges to change the response of gene circuits while lowering protein toxicity [79]. Another approach is to use evolution itself to repair broken synthetic gene circuit components, resulting in more robust gene circuits [31]. Overall, as the library of biological parts increases, the discovery of new “BioBricks” (standardized and interchangeable gene circuits components) will aid in resolving the integrity issues presently associated with synthetic gene circuits. Genomic mining is a promising approach for discovering BioBricks, including identifying novel TetR-family regulators from prokaryotic genomes [80] and CRISPR-Cas systems in microbes [81].

Clinical applications of synthetic gene circuits will continue to expand and could lead to successful treatments for various diseases, including autoimmune disorders [82] and cancers [83]. CAR-T technologies to fight cancer increasingly include synthetic gene circuits and synthetic intercellular pathways to avoid adverse inflammatory reactions that damage healthy cells and to improve the targeting of cancer cells [83, 84]. Additionally, investigating drug resistance in microbial pathogens will require gene circuits that can be introduced into pathogens, which have native gene networks relevant to drug resistance that are complex and incompletely characterized. Relatedly, increasing complexity in gene circuits remains a challenge and will require multiple orthogonal components [80, 85] as well as more advanced computational methods to predict the dynamics of large-scale, nonlinear networks [86]. Ultimately, improvements in our ability to model, design, and construct synthetic gene circuits will benefit biomedical applications as well as increase our understanding of natural gene networks.

Optogenetic gene circuits allow researchers to utilize the strengths that have been developed through two decades of synthetic biology research, as well as to achieve more precise control of living cells. The use of light as a stimulus enables the single-cell control of gene circuit response, which can complement existing systems to study cell populations. The generation of single-cell data will allow researchers to address questions on how individual cells give rise to population level phenomenon and how neighboring cells affect adjacent or distal cells. Answering such questions will be important for extracting information on biological processes such as tissue development [87], epithelial-to-mesenchymal transition [88], and the effects of the microenvironment on cancer progression [89, 90].

While using optogenetic tools will be important for answering a broad range biological questions and for biomedical applications, challenges remain in terms of the scalability and precision of cellular control. There have been applications of optogenetic technology that address these challenges individually. For instance, the light plate apparatus (LPA) [91] is a simple to construct and inexpensive system that can be adapted and used for scaling light-induced conditions *in vitro*. Additionally technology like digital micromirror devices (DMDs) [92] have been used to control single cells in real time. The LPA technology offers scalability but is currently limited in the precision of single-cell control. The DMD on the other hand offers precise single-cell control but is limited by scalability of conditions that can be controlled in a single experiment. Coupling these two types of tools, or their subsequent technological successors, may allow researchers to maximize scalability and optogenetic gene circuit control.

Acknowledgements

DC was supported by funding from the Government of Canada’s New Frontiers in Research Fund – Exploration grant program (NFRFE-2019-01208) and the

University of Alberta. GB was supported by the National Institutes of Health, NIGMS MIRA Program (R35 GM122561) and by the Laufer Center for Physical and Quantitative Biology. MTG was supported by the National Defense Science and Engineering Graduate Fellowship Program.

Author details

Kevin S. Farquhar[1†], Michael Tyler Guinn[2,3,4†], Gábor Balázsi[2,3] and Daniel A. Charlebois[5,6*]

1 Independent Researcher, Houston, USA

2 The Louis and Beatrice Laufer Center for Physical and Quantitative Biology, Stony Brook University, Stony Brook, USA

3 Department of Biomedical Engineering, Stony Brook University, Stony Brook, USA

4 Renaissance School of Medicine, Stony Brook University, Stony Brook, USA

5 Department of Physics, University of Alberta, Edmonton, Canada

6 Department of Biological Sciences, University of Alberta, Edmonton, Canada

*Address all correspondence to: dcharleb@ualberta.ca

† These authors contributed equally to this work.

References

[1] Sorg RA, Gallay C, Van Maele L, Sirard JC, Veening JW. Synthetic gene-regulatory networks in the opportunistic human pathogen Streptococcus pneumoniae. Proc Natl Acad Sci U S A. 2020;117(44): 27608-27619.

[2] Angelici B, Mailand E, Haefliger B, Benenson Y. Synthetic biology platform for sensing and integrating endogenous transcriptional inputs in mammalian cells. Cell Rep. 2016;16(9):2525-2537.

[3] Siu Y, Fenno J, Lindle JM, Dunlop MJ. Design and selection of a synthetic feedback loop for optimizing biofuel tolerance. ACS Synth Biol. 2018; 7(1):16-23.

[4] Honjo H, Iwasaki K, Soma Y, Tsuruno K, Hamada H, Hanai T. Synthetic microbial consortium with specific roles designated by genetic circuits for cooperative chemical production. Metab Eng. 2019;55: 268-275.

[5] Daniel R, Rubens JR, Sarpeshkar R, Lu TK. Synthetic analog computation in living cells. Nature. 2013;497(7451): 619-623.

[6] Huang H, Liu Y, Liao W, Cao Y, Liu Q, Guo Y, et al. Oncolytic adenovirus programmed by synthetic gene circuit for cancer immunotherapy. Nat Commun. 2019;10(1):4801.

[7] Shimizu-Sato S, Huq E, Tepperman JM, Quail PH. A light-switchable gene promoter system. Nat Biotechnol. 2002;20(10):1041-1044.

[8] Gossen M, Bujard H. Tight control of gene expression in mammalian cells by tetracycline-responsive promoters. Proc Natl Acad Sci U S A. 1992;89(12): 5547-5551.

[9] Hasty J, McMillen D, Collins JJ. Engineered gene circuits. Nature. 2002; 420(6912):224-230.

[10] Kaern M, Elston TC, Blake WJ, Collins JJ. Stochasticity in gene expression: From theories to phenotypes. Nature Reviews Genetics. 2005;6(6):451-464.

[11] Fraser D, Kaern M. A chance at survival: Gene expression noise and phenotypic diversification strategies. Molecular Microbiology. 2009;71(6): 1333-1340.

[12] Farquhar KS, Koohi SR, Charlebois DA. Does transcriptional heterogeneity facilitate the development of genetic drug resistance? BioEssays. 2021:e2100043.

[13] Brock A, Chang H, Huang S. Non-genetic heterogeneity – A mutation-independent driving force for the somatic evolution of tumours. Nature Reviews Genetics. 2009;10(5):336-342.

[14] Farquhar KS, Flohr H, Charlebois DA. Advancing antimicrobial resistance research through quantitative Modeling and synthetic biology. Frontiers in Bioengineering and Biotechnology. 2020;8:583415.

[15] McAdams HH, Shapiro L. Circuit simulation of genetic networks. Science. 1995;269(5224):650-656.

[16] Becskei A, Serrano L. Engineering stability in gene networks by autoregulation. Nature. 2000;405 (6786):590-593.

[17] Gardner TS, Cantor CR, Collins JJ. Construction of a genetic toggle switch in Escherichia coli. Nature. 2000;403 (6767):339-342.

[18] Weisberg RA. A genetic switch: Phage lambda and higher organisms. Mark Ptashne. The Quarterly Review of Biology. 1994;69(2):267-268.

[19] Elowitz MB, Leibler S. A synthetic oscillatory network of transcriptional

regulators. Nature. 2000;403(6767): 335-338.

[20] Weber W, Schoenmakers R, Keller B, Gitzinger M, Grau T, Daoud-El Baba M, et al. A synthetic mammalian gene circuit reveals antituberculosis compounds. Proc Natl Acad Sci U S A. 2008;105(29):9994-9998.

[21] Jancik S, Drabek J, Radzioch D, Hajduch M. Clinical relevance of KRAS in human cancers. J Biomed Biotechnol. 2010;2010:150960.

[22] Chen S, Weitemier AZ, Zeng X, He L, Wang X, Tao Y, et al. Near-infrared deep brain stimulation via upconversion nanoparticle-mediated optogenetics. Science. 2018;359(6376): 679-684.

[23] Eickelbeck D, Rudack T, Tennigkeit SA, Surdin T, Karapinar R, Schwitalla JC, et al. Lamprey Parapinopsin ("UVLamP"): A Bistable UV-sensitive Optogenetic switch for ultrafast control of GPCR pathways. ChemBioChem. 2020;21(5): 612–617.

[24] Becskei A, Séraphin B, Serrano L. Positive feedback in eukaryotic gene networks: Cell differentiation by graded to binary response conversion. EMBO J. 2001;20(10):2528–2535.

[25] Nevozhay D, Adams RM, Van Itallie E, Bennett MR, Balazsi G. Mapping the environmental fitness landscape of a synthetic gene circuit. PLoS Computational Biology. 2012;8(4): e1002480.

[26] Alon U. Network motifs: Theory and experimental approaches. Nat Rev Genet. 2007;8(6):450-461.

[27] Camellato B, Roney IJ, Azizi A, Charlebois D, Kaern M. Engineered gene networks enable non-genetic drug resistance and enhanced cellular robustness. Engineering Biology. 2019;3 (4):72-79.

[28] Nevozhay D, Adams RM, Van Itallie E, Bennett MR, Balazsi G. Mapping the environmental fitness landscape of a synthetic gene circuit. PLoS Comput Biol. 2012;8(4):e1002480.

[29] Nevozhay D, Adams RM, Murphy KF, Josic K, Balázsi G. Negative autoregulation linearizes the dose-response and suppresses the heterogeneity of gene expression. Proc Natl Acad Sci USA. 2009;106:5123-5128.

[30] Baron U, Gossen M, Bujard H. Tetracycline-controlled transcription in eukaryotes: Novel transactivators with graded transactivation potential. Nucleic Acids Res. 1997;25(14): 2723-2729.

[31] Gouda MK, Manhart M, Balázsi G. Evolutionary regain of lost gene circuit function. Proc Natl Acad Sci USA. 2019; 116(50):25162-25171.

[32] Gonzalez C, Ray JC, Manhart M, Adams RM, Nevozhay D, Morozov AV, et al. Stress-response balance drives the evolution of a network module and its host genome. Mol Syst Biol. 2015;11(8): 827.

[33] Charlebois DA, Balazsi G, Kaern M. Coherent feedforward transcriptional regulatory motifs enhance drug resistance. Physical Review E. 2014;89 (5):052708.

[34] Balzi E, Goffeau A. Yeast multidrug resistance: The PDR network. J Bioenerg Biomembr. 1995;27(1):71-76.

[35] Misra S, Ghatak S, Toole B. Regulation of MDR1 expression and drug resistance by a positive feedback loop involving hyaluronan, phosphoinositide 3-kinase, and ErbB2. J Biol Chem. 2005;280:20310-20315.

[36] Gillespie DT. A general method for numerically simulating the stochastic time evolution of coupled chemical reactions. J Comput Phys. 1976;22: 403-434.

[37] Gillespie DT. Exact stochastic simulation of coupled chemical reactions. J Phys Chem. 1977;81: 2340-2361.

[38] Sambrook J. Molecular cloning: A laboratory manual: Third edition. Cold Spring Harbor, N.Y.: Cold Spring Harbor Laboratory Press [2001] ©2001; 2001.

[39] Farquhar KS, Charlebois DA, Szenk M, Cohen J, Nevozhay D, Balazsi G. Role of network-mediated stochasticity in mammalian drug resistance. Nat Commun. 2019;10(1):2766.

[40] Nevozhay D, Zal T, Balazsi G. Transferring a synthetic gene circuit from yeast to mammalian cells. Nat Commun. 2013;4:1451.

[41] Guinn M, Bleris L. Biological 2-input decoder circuit in human cells. ACS Synth Biol. 2014;3(8):627-633.

[42] Bleris L, Xie Z, Glass D, Adadey A, Sontag E, Benenson Y. Synthetic incoherent feedforward circuits show adaptation to the amount of their genetic template. Mol Syst Biol. 2011;7:519.

[43] Auslander S, Stucheli P, Rehm C, Auslander D, Hartig JS, Fussenegger M. A general design strategy for protein-responsive riboswitches in mammalian cells. Nat Methods. 2014;11(11): 1154-1160.

[44] Thibodeaux GN, Cowmeadow R, Umeda A, Zhang Z. A tetracycline repressor-based mammalian two-hybrid system to detect protein-protein interactions in vivo. Anal Biochem. 2009;386(1):129-131.

[45] Kramer BP, Fussenegger M. Hysteresis in a synthetic mammalian gene network. Proc Natl Acad Sci U S A. 2005;102(27):9517-9522.

[46] Bacchus W, Lang M, El-Baba MD, Weber W, Stelling J, Fussenegger M. Synthetic two-way communication between mammalian cells. Nat Biotechnol. 2012;30(10):991-996.

[47] Boyden ES, Zhang F, Bamberg E, Nagel G, Deisseroth K. Millisecond-timescale, genetically targeted optical control of neural activity. Nat Neurosci. 2005;8(9):1263-1268.

[48] Zoltowski BD, Crane BR. Light activation of the LOV protein vivid generates a rapidly exchanging dimer. Biochemistry. 2008;47(27):7012-7019.

[49] Wang X, Chen X, Yang Y. Spatiotemporal control of gene expression by a light-switchable transgene system. Nat Methods. 2012;9 (3):266-269.

[50] Muller K, Engesser R, Metzger S, Schulz S, Kampf MM, Busacker M, et al. A red/far-red light-responsive bi-stable toggle switch to control gene expression in mammalian cells. Nucleic Acids Res. 2013;41(7):e77.

[51] Levskaya A, Weiner OD, Lim WA, Voigt CA. Spatiotemporal control of cell signalling using a light-switchable protein interaction. Nature. 2009;461 (7266):997-1001.

[52] Renicke C, Schuster D, Usherenko S, Essen LO, Taxis C. A LOV2 domain-based optogenetic tool to control protein degradation and cellular function. Chem Biol. 2013;20(4):619-626.

[53] Ye H, Daoud-El Baba M, Peng RW, Fussenegger M. a synthetic optogenetic transcription device enhances blood-glucose homeostasis in mice. Science. 2011;332(6037):1565-1568.

[54] Folcher M, Oesterle S, Zwicky K, Thekkottil T, Heymoz J, Hohmann M, et al. Mind-controlled transgene expression by a wireless-powered optogenetic designer cell implant. Nat Commun. 2014;5:5392.

[55] Guinn MT. Engineering Human Cells with Synthetic Gene Circuits

Elucidates How Protein Levels Generate Phenotypic Landscapes [Ph.D. Thesis]. Ann Arbor: State University of New York at Stony Brook; 2020.

[56] Zhao W, Bonem M, McWhite C, Silberg JJ, Segatori L. Sensitive detection of proteasomal activation using the Deg-on mammalian synthetic gene circuit. Nat Commun. 2014;5:3612.

[57] Charlebois DA, Diao J, Nevozhay D, Balazsi G. Negative regulation gene circuits for efflux pump control. Methods Mol Biol. 2018;1772:25-43.

[58] May T, Eccleston L, Herrmann S, Hauser H, Goncalves J, Wirth D. Bimodal and hysteretic expression in mammalian cells from a synthetic gene circuit. PLoS One. 2008;3(6):e2372.

[59] Li Y, Moore R, Guinn M, Bleris L. Transcription activator-like effector hybrids for conditional control and rewiring of chromosomal transgene expression. Sci Rep. 2012;2:897.

[60] Guinn MT, Balazsi G. Noise-reducing optogenetic negative-feedback gene circuits in human cells. Nucleic Acids Res. 2019;47(14):7703-14.

[61] Lee D, Hyun JH, Jung K, Hannan P, Kwon HB. A calcium- and light-gated switch to induce gene expression in activated neurons. Nat Biotechnol. 2017; 35(9):858-863.

[62] Polstein LR, Gersbach CA. A light-inducible CRISPR-Cas9 system for control of endogenous gene activation. Nat Chem Biol. 2015;11(3):198-200.

[63] Ma Z, Du Z, Chen X, Wang X, Yang Y. Fine tuning the LightOn light-switchable transgene expression system. Biochem Biophys Res Commun. 2013; 440(3):419-423.

[64] Chen X, Wang X, Du Z, Ma Z, Yang Y. Spatiotemporal control of gene expression in mammalian cells and in mice using the LightOn system. Curr Protoc Chem Biol. 2013;5(2):111-129.

[65] Olson EJ, Hartsough LA, Landry BP, Shroff R, Tabor JJ. Characterizing bacterial gene circuit dynamics with optically programmed gene expression signals. Nat Methods. 2014;11(4): 449-455.

[66] Bradley RW, Buck M, Wang B. Tools and principles for microbial gene circuit engineering. J Mol Biol. 2016;428 (5 Pt B):862-888.

[67] Sadat Mousavi P, Smith SJ, Chen JB, Karlikow M, Tinafar A, Robinson C, et al. A multiplexed, electrochemical interface for gene-circuit-based sensors. Nat Chem. 2020;12(1):48-55.

[68] Harris SL, Levine AJ. The p53 pathway: Positive and negative feedback loops. Oncogene. 2005;24(17): 2899-2908.

[69] Yu P, Kosco-Vilbois M, Richards M, Kohler G, Lamers MC. Negative feedback regulation of IgE synthesis by murine CD23. Nature. 1994;369(6483): 753-756.

[70] Deans TL, Cantor CR, Collins JJ. A tunable genetic switch based on RNAi and repressor proteins for regulating gene expression in mammalian cells. Cell. 2007;130(2):363-372.

[71] Madar D, Dekel E, Bren A, Alon U. Negative auto-regulation increases the input dynamic-range of the arabinose system of Escherichia coli. BMC Syst Biol. 2011;5:111.

[72] Gossen M, Freundlieb S, Bender G, Muller G, Hillen W, Bujard H. Transcriptional activation by tetracyclines in mammalian cells. Science. 1995;268(5218):1766-1769.

[73] Forster K, Helbl V, Lederer T, Urlinger S, Wittenburg N, Hillen W. Tetracycline-inducible expression systems with reduced basal activity in

mammalian cells. Nucleic Acids Res. 1999;27(2):708-710.

[74] Muller K, Zurbriggen MD, Weber W. An optogenetic upgrade for the Tet-OFF system. Biotechnol Bioeng. 2015;112(7):1483-1487.

[75] Usherenko S, Stibbe H, Musco M, Essen LO, Kostina EA, Taxis C. Photo-sensitive degron variants for tuning protein stability by light. BMC Syst Biol. 2014;8:128.

[76] Klotzsche M, Berens C, Hillen W. A peptide triggers allostery in tet repressor by binding to a unique site. J Biol Chem. 2005;280(26):24591-24599.

[77] Sakai S, Ueno K, Ishizuka T, Yawo H. Parallel and patterned optogenetic manipulation of neurons in the brain slice using a DMD-based projector. Neurosci Res. 2013;75(1): 59-64.

[78] Rullan M, Benzinger D, Schmidt GW, Milias-Argeitis A, Khammash M. An Optogenetic platform for real-time, single-cell interrogation of stochastic transcriptional regulation. Mol Cell. 2018;70(4):745-756 e6.

[79] Wan X, Pinto F, Yu L, Wang B. Synthetic protein-binding DNA sponge as a tool to tune gene expression and mitigate protein toxicity. Nature Communications. 2020;11(1): 5961.

[80] Stanton BC, K NAA, Tamsir A, Clancy K, Peterson T, Voigt CA. Genomic mining of prokaryotic repressors for orthogonal logic gates. Nature Chemical Biology. 2014;10(2): 99-105.

[81] Burstein D, Harrington LB, Strutt SC, Probst AJ, Anantharaman K, Thomas BC, et al. New CRISPR-Cas systems from uncultivated microbes. Nature 2017;542(7640):237-241.

[82] Smole A, Lainscek D, Bezeljak U, Horvat S, Jerala R. A synthetic mammalian therapeutic gene circuit for sensing and suppressing inflammation. Mol Ther. 2017;25(1):102-119.

[83] Nissim L, Wu MR, Pery E, Binder-Nissim A, Suzuki HI, Stupp D, et al. Synthetic RNA-Based Immunomodulatory Gene Circuits for Cancer Immunotherapy. Cell. 2017;171 (5):1138-50 e15.

[84] Choe JH, Watchmaker PB, Simic MS, Gilbert RD, Li AW, Krasnow NA, et al. SynNotch-CAR T cells overcome challenges of specificity, heterogeneity, and persistence in treating glioblastoma. Sci Transl Med. 2021;13(591).

[85] Szenk M, Yim T, Balázsi G. Multiplexed gene expression tuning with orthogonal synthetic gene circuits. ACS Synth Biol. 2020;9(4):930-939.

[86] Bashor CJ, Patel N, Choubey S, Beyzavi A, Kondev J, Collins JJ, et al. Complex signal processing in synthetic gene circuits using cooperative regulatory assemblies. Science. 2019;364(6440):593.

[87] Mayr U, Serra D, Liberali P. Exploring single cells in space and time during tissue development, homeostasis and regeneration. Development. 2019; 146(12).

[88] Li C, Balazsi G. a landscape view on the interplay between EMT and cancer metastasis. NPJ Syst Biol Appl. 2018;4:34.

[89] Horsman MR, Vaupel P. Pathophysiological basis for the formation of the tumor microenvironment. Front Oncol. 2016;6:66.

[90] Bailey PC, Lee RM, Vitolo MI, Pratt SJP, Ory E, Chakrabarti K, et al. Single-Cell Tracking of Breast Cancer Cells Enables Prediction of Sphere Formation from Early Cell Divisions. iScience. 2018;8:29-39.

[91] Gerhardt KP, Olson EJ, Castillo-Hair SM, Hartsough LA, Landry BP, Ekness F, et al. An open-hardware platform for optogenetics and photobiology. Sci Rep. 2016;6:35363.

[92] Milias-Argeitis A, Summers S, Stewart-Ornstein J, Zuleta I, Pincus D, El-Samad H, et al. In silico feedback for in vivo regulation of a gene expression circuit. Nat Biotechnol. 2011;29(12): 1114-1116.